BIM and Construction Management

BIM and Construction Management

Proven Tools, Methods, and Workflows

Brad Hardin

WILEY

Wiley Publishing, Inc.

Senior Acquisitions Editor: WILLEM KNIBBE
Development Editor: DICK MARGULIS
Technical Editor: MARTIN VIVEROS
Production Editor: RACHEL MCCONLOGUE
Copy Editor: KIM WIMPSETT
Production Manager: TIM TATE
Vice President and Executive Group Publisher: RICHARD SWADLEY
Vice President and Publisher: NEIL EDDE
Book Designer: FRANZ BAUMHACKL
Compositor: CHRIS GILLESPIE, HAPPENSTANCE TYPE-O-RAMA
Proofreader: NANCY BELL
Indexer: JACK LEWIS
Project Coordinator, Cover: LYNSEY STANFORD
Cover Designer: RYAN SNEED
Cover Image: ZEN SHUI/JAMES HARDY/PHOTOALTO AGENCY/JUPITER IMAGES

For general information on our other products and services or to obtain technical support, please contact our Customer Care Department within the U.S. at (877) 762-2974, outside the U.S. at (317) 572-3993 or fax (317) 572-4002.

Wiley also publishes its books in a variety of electronic formats. Some content that appears in print may not be available in electronic books.

Library of Congress Cataloging-in-Publication Data

Hardin, Brad, 1980-

BIM and construction management : proven tools, methods, and workflows / Brad Hardin. — 1st ed.

p. cm.

ISBN-13: 978-0-470-40235-1 (pbk.)

ISBN-10: 0-470-40235-0

1. Building information modeling. I. Title.

TH437.H287 2009

690.0285—dc22

2009004123

10 9 8 7 6 5 4 3 2 1

Dear Reader,

Thank you for choosing *BIM and Construction Management: Proven Tools, Methods, and Workflows*. This book is part of a family of premium-quality Sybex books, all of which are written by outstanding authors who combine practical experience with a gift for teaching.

Sybex was founded in 1976. More than 30 years later, we're still committed to producing consistently exceptional books. With each of our titles we're working hard to set a new standard for the industry. From the paper we print on to the authors we work with, our goal is to bring you the best books available.

I hope you see all that reflected in these pages. I'd be very interested to hear your comments and get your feedback on how we're doing. Feel free to let me know what you think about this or any other Sybex book by sending me an email at nedde@wiley.com, or if you think you've found a technical error in this book, please visit http://sybex.custhelp.com. Customer feedback is critical to our efforts at Sybex.

Best regards,

Neil Edde
Vice President and Publisher
Sybex, an imprint of Wiley

To my Grandfather, Sy Hardin, whom I love and respect.

 # Acknowledgments

First, I would like to thank my wife Iris and daughter Lucia for the amount of time they shared me with the creation of this book, the tolerance of late nights, and without whose support this book would not have been possible. I consider myself blessed.

Additionally, I would like to thank my colleagues and fellow "BIMers" who always have great input, experiences, and perspectives to share. A special thanks to Brian Arnold, MGC, my partners from the Virtual Construct lab team and Martin Viveros, technical editor extraordinaire, who helped look over my shoulder and make every tutorial accurate. As always, my thanks go to Eddy Krygiel for his insight, expertise, and the answer to the occasional "How has it really worked for you?" question. I would also like to thank the AIA, FMI, AGC, and the AISC members who are too numerous to name, but who provided volumes of invaluable information. A special thanks to Erika Winters-Downey for her insight and contribution.

I would also like to thank Robert Forest at Smith and Gill for providing the Masdar City Case Study, and who found time in between trips and project management to help out. Additionally, thanks to Tim Case from Parsons-Brinckerhoff who provided the Oakland Bridge Case Study and who showed just how powerful BIM and graphics can be.

A big thanks to all of the people from Wiley: Willem Knibbe, Dick Margulis, Rachel McConlogue, Pete Gaughan, and Kim Wimpsett for all of their help, patience, encouragement, prodding, and guidance. I truly could not have accomplished this without your help and am grateful to all of you.

Many other architects, engineers, professors, students, and contractors along the way have helped me in my efforts and I sincerely thank all of you for your dialog, concerns, and desire for a book that shows "how" to use BIM for construction management. I hope you are not disappointed.

Lastly, I would like to thank the BIM community at large, for their effort to push the boundaries of integration, technology, and enthusiasm about using BIM to affect our industry for the better.

About the Author

Brad Hardin, AIA, LEED AP, is a founding partner of Virtual Construct lab, a BIM consulting company. He has been responsible for the successful integration of BIM into multiple organizations and has trained, lectured, and provided consulting to owners, government organizations, AEC teams, and manufacturing companies. Brad has co-authored several papers and continues to be actively involved in the BIM design and construction community.

Contents

Foreword

I'm what you'd call an "early adopter" of BIM. I started using Autodesk's BIM product, Revit, back in 2001 because (as one of the principals of my firm at the time put it), "We'd just landed a project bigger than anything we'd ever done and we felt we needed a new manner of looking at documentation if we were going to succeed." I've seen and heard that same sentiment a lot over the years as BIM has caught on like wildfire within the AEC industry. I've watched as many firms have adopted BIM as a new way of documenting and designing, and in the process created a new culture within their firms as a byproduct of BIM integration. This change in workflow towards more design visualization, better building metrics, building analysis and more is changing the way we communicate and design.

We are no longer trading two dimensional ideas and concepts on paper with our owners, consultants, or contractors – we're trading in three-dimensional virtualized buildings that contain vast amounts of useful information. This virtualized world allows us an unprecedented amount of control and knowledge over the building before we even put a shovel in the ground.

And yet BIM is still in its infancy, and the process of BIM is really one of evolution, not revolution. You will never be able to achieve all of the amazing things possible with BIM on your very first project. Or your second. Or possibly even your third. Like design, it is a process of continual improvement – not only in modeling, but in sharing that data and information with team members. The future of BIM, and our willingness to learn from other team members, can help us move quickly to a better future. There is no future, no next, if we do not change the ways in which we work, live, and play. If we are open to change, then a few things are inevitable.

Parametric modeling will go well beyond mapping relationships between objects and assemblies. Both model and designer will have knowledge of climate and region. The model will know its building type, insulation values, solar heat gain coefficients, and structural components. It will inform the design team with regard to materials, locations, and costs. As the building is modeled, the designer will instantly see the impact of the building orientation and envelope choices on the sizing of the mechanical system. It will analyze the design for Americans with Disabilities Act (ADA) compliance and code-related issues. It will be a system completely interactive with key building information, so that design integration and data return among all systems is immediate. And after the building is occupied, it will create an opportunity for a post-occupancy and building life-cycle feedback loop.

However, BIM will not be the solution unto itself. The solution will continue to rely on our abilities to use the tool to its highest advantage. Through the use of BIM, we are able to move from a documentation system that is fragmented and inherently unintelligent to one that is centrally based and able to parametrically analyze model data almost instantly. In our legacy system, individual drawings and lines have no value other than their printed form. With BIM, the intelligence can be interlinked between objects within the modeled assemblies, allowing our team members to be both specialized and integrated. With BIM, the design process requires team communication and integration—with *all* the team members.

Many of the BIM books out in the market right now are solely focused on BIM from the view of a designer. In order to get the most out of this new process, designers need to understand BIM from the point of view of the *builder*. Builders in turn, need to understand the best ways to leverage and incorporate this paradigm-shifting technology and become equal partners in the conversation. What Brad has accomplished in this book is not a revolution; it is the simple evolution to an idea that has been cooking within our industry for the past decade. That powerful idea is communication and the sharing of information with all the project stake holders for the greater good of the project. And what Brad has done is given a clear view of a critical piece to a BIM solution from the builder's perspective. Full of incredibly useful field-tested techniques, tips, methods, and real-world workflows, this is the ultimate resource for builders and others in the construction industry needing to implement BIM, and it will help both builders and designers better communicate and collaborate. Because at the end of the day, it's not about the designer or the builder, it's about the project. BIM allows us to get there in a way we've never been able to do before.

Enjoy.

EDDY KRYGIEL
HNTB
Sr. Project Architect
BIM + Sustainability Specialist
co author Introducing Revit 2010, Mastering Revit 2010, *Green BIM*

Introduction

This book was created for those who are involved in the building information modeling (BIM) community or who want to learn more about BIM and what it means for construction management professionals. This book introduces what BIM is conceptually and shows the real-world ways BIM technologies can be put to use. BIM represents a fundamental change in the way design and construction professionals work, and the BIM opportunities that are currently being leveraged and explored by construction companies around the globe are staggering. This book represents years of study, practice, and discussion of BIM, how it functions in current methods, and what is being changed to better enable this technology in the design and construction industries. I hope this book will be one of the building blocks, along with others' experience, that can turn BIM into a more refined and developed process.

How to Use This Book

As you will hear multiple times throughout these pages, this book is for those who want to define a process using BIM. The chapters show at what points in a design and construction process certain tasks need to be accomplished to achieve a certain result, and the chapters often include tutorials showing how to accomplish that task using specific software. By completing the step-by-step tutorials in this book, you'll gain a perspective of the time and effort it takes to achieve the desired product. In addition, you'll increase your working knowledge of the systems involved. This book also will outline how and to whom the completed information should be distributed and how to archive and maintain a history of information for a project. I'll show how to quantify the users' individual results along the way by analyzing savings, costs, and time compared to a 2D CAD process. By delving into the process and then laying a foundation of metrics to back up the effort, a BIM user can see what tools might be useful as they begin different projects with varying degrees of complexity.

Goals

BIM and Construction Management has two goals:

- To show the tools and resources that the design and construction communities are using to manage information, with examples of how team members are contributing to the project by streamlining their day-to-day practices and making their efforts more effective

- To outline processes that construction professionals can expand upon as BIM technology grows and develops

Figure I.1 show the process and tools I will use in this book along a construction process. The tools and processes outlined in this book are associated with a design-build/integrated method of delivery, but keep in mind that there are many other methods of delivery. Although BIM can fit into older practices, as this book will show, BIM is most effectively used in collaborating with integrated teams.

Also, this book makes reference to the *CAD process*. This is for brevity's sake; please understand that this refers to a 2D line/vector–based means of creating and managing documentation throughout a construction project. In turn, a *BIM process* refers to methods that enable BIM models as well as tasks that correlate to pieces of a CAD process but are different. Although computer-aided design (CAD) is frequently used in the industry, it is rarely used when in reference to BIM technology. For this reason, CAD process and CAD technology throughout this book will refer to the use of vector-based documentation as opposed to virtual constructions and models.

Software and Tutorials

The information in this book is based on case studies, interviews with colleagues, and my own industry experiences. Although this book describes several key software systems and gives tutorials on how to use these tools, it is not my intention to recommend any specific software. Many software products can do the tasks described in this book in similar if not exact ways. Moreover, it is not my intent to say that using this software is the right or only way of accomplishing BIM-related tasks. Rather, this book uses the tools, as well as the processes associated with them, that my colleagues and I are most familiar with using.

This book identifies a process of managing, testing, and extracting valuable data during design and construction. Although many professionals have yet to see the value in BIM, it is my purpose to make a case for BIM and to outline a set of goals, timelines, proven lessons learned, and best-practice scenarios from experience. These valuable lessons to the AECO team call for more and more integration with design professionals to achieve extraordinary results.

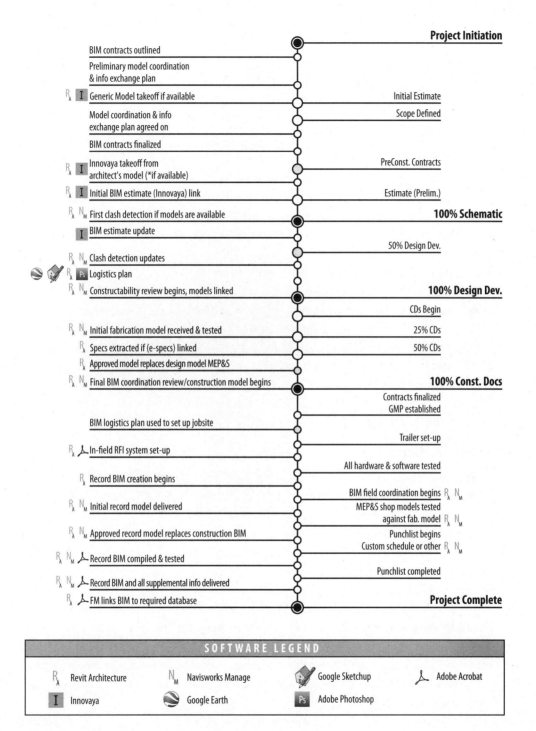

Figure I.1 The BIM process outlined in this book during a construction project

This book will be valued by the user base, or *doers*, intimately involved with integrating, managing, and growing BIM processes in the field of construction management. This book does not aim to be a purely academic discussion, although I discuss the future of BIM in the last chapter of this book and speculate on exciting opportunities this technology could offer. Rather, it identifies deliverables along a construction timeline that, when introduced into a construction process, show how team performance and coordination can be vastly improved. This book uses tutorials to educate construction managers about the time, tools, and resources needed to complete each task. From these tutorials, construction managers can make more informed decisions about implementing a BIM process and managing schedules as well as standards. It is my aim that this book be used as a resource and a how-to guide for BIM users and students.

The tutorials in this book range from beginner user to advanced in order to challenge users and students alike to better understand the technology necessary to accomplish BIM. By introducing the tools as concepts first and then showing by example, this book equips students and new users with a fundamental understanding first of the task to be accomplished, then why it is important, and finally how to accomplish it. BIM is a new and exciting way of engaging multiple stakeholders, educating new associates, and creating value that has just begun to take hold in the design and construction industries.

What's Inside

Chapter 1, "BIM and Construction Management," makes the case for BIM. I explore how BIM is useful to contractors, subcontractors, and construction administrators and show how this software and technology is not a passing fad. Chapter 1 reviews the tools and methods used today and then outlines a plan to implement BIM into a company's infrastructure.

Chapter 2, "BIM and Preconstruction," takes an in-depth look at the services, tools, and workflows in BIM at the preconstruction level. It includes tutorials on how to create cost estimates, how to move models and quantities between programs, and how to generate a simple constructability issue report that is compiled by the BIM user. Chapter 2 demonstrates schedule linking to a model and shows the value of providing a three-dimensional tool that clearly describes to the construction team what components are to be erected and finished along the construction manager's timeline.

Chapter 3, "BIM and Construction," deals specifically with the relationship between the contractor and the design team's model. It focuses on goals and scheduling, constructability model reviews, and trade coordination. Chapter 3 gives tutorials using Navisworks to update a schedule animation, to prepare the file for future schedule alterations, and to generate a clash detection report.

Chapter 4, "BIM and Updates," looks at BIM and updates. This chapter outlines how to set up a file management strategy that creates subsequent efficiencies. It also shows how to update a model or the associated information during prebid, estimating, and model updates during design.

Chapter 5, "BIM and Construction Administration," shows how BIM is used in the field. The tasks outlined include the job trailer, the superintendent's role, BIM and RFIs, BIM and field coordination, BIM punch lists, and project closeout. This chapter focuses on how BIM technology is applied in the real world and for what purposes it's used during construction. I also compare the traditional CAD-based methods of field use vs. a BIM project. Finally, I show how to prepare the record BIM for delivery to the owner.

Chapter 6, "BIM and Sustainability," examines the contractor's new role in a changing market and what advantages BIM offers. This chapter explains some of the global economic opportunities regarding BIM and green practice. We look at the tools available to the builder for sustainable analysis. The chapter shows how to use the BIM model to generate custom material reuse reports using Revit schedules and the quantities generator and how to produce reports that show the weight, cost, and amount of reused materials in a project. The last portion of Chapter 6 outlines some of the tools available to architects and engineers to test the complete BIM model to achieve faster calculations and decrease the amount of time they must invest in doing calculations to inform the team of results.

Chapter 7, "BIM and Facility Management," explores an area largely undefined yet growing in demand. It defines the current challenges owners and facility managers face in rising energy and equipment costs, information management, and outdated drawings. This chapter also discusses life cycle information management. Additionally, it shows the potential for BIM to be used after construction when the model is transferred to the owner.

Chapter 8, "The Future of BIM," explains the current trends in the industry as well as the future opportunities in using BIM. This chapter explores the potential multiple and simultaneous testing of models as the technology continues to be refined. It also speculates as to software not yet in the market that could influence much of the way design and construction professionals practice. The chapter concludes with thoughts on where BIM can take us and how we must achieve better communication and collaboration for our industry to thrive.

Conclusion

Although it is the aim of this book to show a method of using BIM, it would be foolish to say that the suggestions in this book are the only solutions available today. Even as this book's ink is being put to paper, there are probably new and improved technologies that are entering the marketplace that could better facilitate a portion of work to be completed. Creating a strong foundation of a process, which can continually be refined, adds clarity to the construction manager's role in BIM. In BIM, the model is much more than just a virtual representation of a building. Ultimately, we will find how construction management *is* information management.

How to Contact the Author

I welcome feedback from you about this book or about books you'd like to see from me in the future. You can reach me by writing to bhardin@virtualconstructlab.com. For more information about my books and articles, visit my website at www.virtualconstructlab.com.

Sybex strives to keep you supplied with the latest tools and information you need for your work. Please check its website at www.sybex.com, where we'll post additional content and updates that supplement this book should the need arise. Enter **BIM** in the Search box (or type the book's ISBN—**9780470402351**), and click Go to get to the book's update page.

BIM and Construction Management

This chapter introduces building information modeling (BIM). It discusses why BIM is becoming the industry standard and how it is transforming the global construction community at large. This chapter defines a new way of thinking about BIM as a process, not just as software, and specifically explores the following topics:

1

In this chapter

The value and potential of BIM technology

Existing delivery methods

A new concept of delivery

A new concept of process

The ten steps for successful implementation

The Value and Potential of BIM Technology

BIM is a revolutionary technology and process that has transformed the way buildings are designed, analyzed, constructed, and managed. Currently, an overwhelming amount of information is available about BIM, such as theories on where BIM can go, the vast array of tools available, and how BIM seems to be the answer to all the problems facing a construction manager (CM). Although some of this information is useful, often it inundates potential users because the information all seems to meld together. BIM has become a proven technology. What it can do and the concepts associated with BIM taken out of context, however, can become misleading and frustrate users and owners alike to the point of not wanting to use this technology again on future projects. This not only hurts the future growth of BIM technology, but it inhibits users from getting involved and sharing their experiences with others in the BIM community to further refine lessons learned and best practices. Figure 1.1 shows an example of a building constructed using BIM technology.

COURTESY OF McCOWNGORDON CONSTRUCTION, LLC.

Figure 1.1 Sunset Drive office building, a LEED Gold building constructed using BIM technology

BIM works. While there currently are a number of inefficiencies that will continue to be refined, BIM as a technology is no longer in its infancy and has started to produce results for the AEC/O industry all over the world. The new frontier for BIM and for its users is to define a new process that better enables this new technology. This book identifies a new process and a way of thinking about BIM that is different than previous processes based off older technology.

BIM: A Primer

So, what is BIM? As Charles Eastman puts it in *Building Product Models: Computer Environments Supporting Design and Construction* (CRC Press, 1999), "BIM is a digital representation of the building process to facilitate exchange and interoperability of information in digital format." For a contractor, BIM is the virtual construction of a facility or structure that contains intelligent objects in a single source file that, when shared among project team members, intends to increase the amount of communication and collaboration. The words *communication* and *collaboration* have become common in discussions about BIM today, not only among architects, engineers, and contractors but also with owners, facility managers, and sustainable design professionals. In fact, according to *Interoperability in the Construction Industry* (McGraw Hill Construction, 2007), construction productivity has decreased significantly over the last forty years. This is in large part because of a lack of communication and collaboration through information (Figure 1.2).

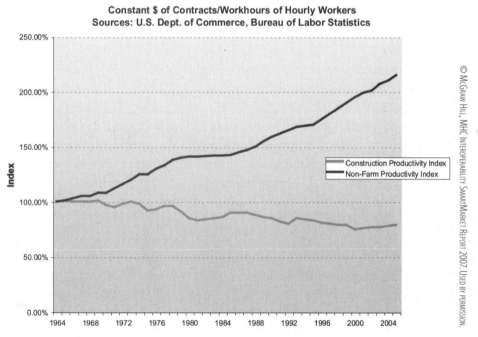

Figure 1.2 Construction productivity index compared to nonfarm industries

Informed contractors and sophisticated owners have begun to look at the current processes and demand higher interoperability among teams and among software packages, better tools, fewer change orders, and fewer questions in the field.

The question then becomes, how? How can building professionals begin to deliver better projects to their owners even as buildings become more and more complex and dependent on new technologies in an ever-changing and moving world? One of the loudest answers has been BIM.

BIM is not just software. BIM is a process *and* software. Many believe that once they have purchased a license for a particular piece of BIM software, they can sit someone in front of the computer and they are now "doing BIM." What many don't realize, though, is that building information modeling means not only using three-dimensional modeling software but also implementing a new way of thinking. It is in essence a new way of *not* doing the same old thing. In my experience, as a company integrates this technology, it begins to see other processes start to change. Where a certain process might have made perfect sense for a CAD-type technology, now that doesn't seem to be as efficient. As the technology changes, so do the practices and functions of the people using the technology. In other words, don't expect to begin adapting to this new technology and have everything function as it has in the past. Chances are that very few of your practices will remain the same, because when the information is much richer and more robust, the management of this information must change in order to fully utilize its potential. Although it is clear that many BIM technologies continue to grow and develop, it is even more apparent that the "old way of doing things" has a limited future.

So, what are the advantages of BIM? Let's first look at the owner's perspective. According to *Interoperability in the Construction Industry*, 49 percent of owners are now demanding BIM be used on their projects (Figure 1.3). Right behind the owners' demand for BIM, 47 percent of construction industry professionals are choosing to use BIM for its "ability to improve communication with clients/others in the design and construction process." Clearly, BIM is being perceived by owners as a tool that can better coordinate and manage building information. Additionally, construction industry professionals are choosing to use BIM to improve the design and construction process. Although the technology is key, it is perhaps even more critical to define processes that utilize this technology and how to work better with all members involved.

BIM and the Team

What does BIM mean to other team members? Architects use it to more efficiently model their designs (it's not drafting anymore), to generate the documents that are required of them, and to perform a host of other tasks. Designers using BIM can quickly generate rendered perspective views and animations to better communicate the project to the owner or local municipalities. Engineers can model mechanical and electrical designs to evaluate how a system will perform. Sustainability consultants, architects, and engineers can measure day lighting, recycled and reused material content, and solar orientation. In essence, any physically modeled object can be created, infused with data, analyzed, scheduled, and tested.

Factors Influencing the Use of BIM

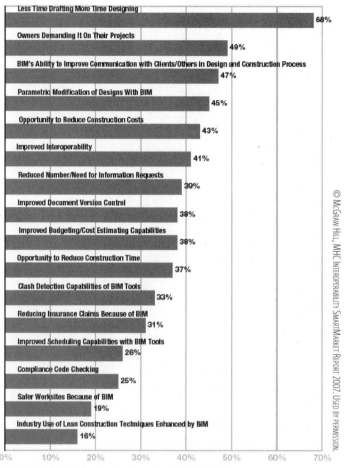

Source: McGraw-Hill Construction Research and Analytics, 2007

Figure 1.3 Industry factors influencing the use of BIM

Existing Delivery Methods

So, why does BIM matter to contractors? To really understand the answer to this question, you need to first look at current processes to see how information is shared, what types of technologies are being used, and what types of project delivery methods are being used. Therefore, in the following sections, I'll introduce you to four current project delivery methods: design-bid-build, design-build, CM-at-risk, and integrated project delivery. I'll talk about each method in the context of five categories, specifically in regards to information and workflow:

•   Preconstruction

•   Communication and collaboration methods

- Types of documents
- Clarification of information
- Project closeout

The four methodologies discussed are practiced using CAD technology. Although the type of project delivery varies for the purposes of this discussion, I will cover the most popular methods and how information flows within each of them. As there are other types of delivery methods, these commonly practiced methods will begin to paint a picture of how information is currently shared among teams and the last method will explore a potential future means of project delivery.

Design-Bid-Build

Design-bid-build is one of the most traditional types of delivery methods practiced today. The basic concept behind design-bid-build is a linear process. The owner contracts with the architect to develop a program and then further develops the design using mechanical, electrical, and plumbing engineers. After the design has been solidified, the project moves into construction documentation, with the understanding that the design team will produce completed construction documents for the project to be issued to a number of general contractors to bid on. The role of the general contractor on this type of delivery is to take the documents and specifications and work with subcontractors to define their relevant scope of work and deliver a bid for the project. Using these subcontractors' estimates, the general contractor then compiles a completed bid. This bid is then delivered to the owner, and at this point all other bids on the project are opened either privately or publicly depending on the project type. The owner then awards the general contractor the project's contract based upon price to complete the project.

The design-bid-build delivery method has these problems:

- If a contractor has been consulted to complete an estimate on the project during construction documentation, the project may go over its scheduled delivery date because of additional drawing time due to value engineering.
- It assumes that cheaper is better. Although that assumption might be correct in regard to cost, it is not necessarily accurate in regard to project quality or the ability of the contractor to adequately perform the work or collaborate with the team.
- In a design-bid-build delivery it is the assumption that by promoting competition among general contractors the best possible price will be issued.
- The owner is at risk to the contractor for design errors.

General contractors' bids on this type of project may vary wildly because of both internal and perceived external issues on the project. First, if a general contractor is backlogged and has too much work on their plate, they might bid the project higher. This contractor wants to complete the work they already have and justify the additional cost through staff adjustment, overtime, and other overhead costs to complete the work. Second, the general contractor will gauge the aptitude of the design team based on the documents. Because this is often the only means of collaboration with the design team that the contractor will have during a bid process, aside from a pre-bid meeting, they will raise or lower their costs depending on the detail and accuracy of the documents. Lastly, the contractor is at risk of not being selected. Typically general contractors spend a considerable amount of time and money on producing a bid, and there is a high risk for not being rewarded for that investment. Additionally, even if they are the low bidder on the project, the owner reserves the right to not accept any of the bids regardless of the cost. This drives some contractors to work with owners under other delivery methods that validate their investment of resources to receive a project.

Preconstruction

In a design-bid-build contract (Figure 1.4), typically no information is shared in schematic design (SD) or design development (DD) between the architect and contractor. Although a contractor might be involved with a design-bid-build project as an owner's representative or in a design-assist capacity, often that contractor is involved purely for estimate checking and cannot bid on the project, because they might have additional information that would be an advantage over the other bidders and because their contract is for a predetermined fee separate from the construction contract. Since little to no contractor involvement during preconstruction severely limits the design team's ability to make informed design decisions, the design team is forced to issue "value engineering" options or "deduct" options to reduce the bid amounts for the project.

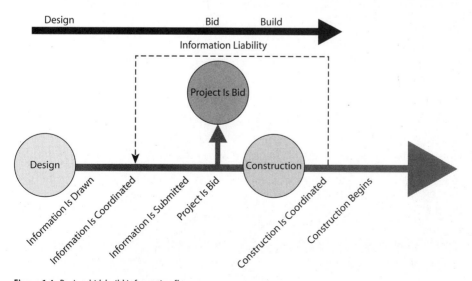

Figure 1.4 Design-bid-build information flow

Design-bid-build is not all bad, in that it allows the architects and engineers some time to collaborate effectively and produce relatively integrated documents. Design-bid-build also gives the owner control of the project, but requires a high level of owner expertise and resources. In regard to BIM it is mostly ineffective. The design-bid-build delivery method limits the ability for BIM to be used to its full potential as a coordination tool by the contractor. In regard to scheduling, clash detection, constructability, and estimating, the model is somewhat of an afterthought, because the drawings take precedence and because the architects and engineers are under no obligation to even share the model—if it exists—with the contractor. In reality, BIM is not useful in a design-bid-build with the exception of trying to use it to quickly extract quantities for estimating purposes from a model of unknown quality. The attitude for model sharing is sometimes tentative if language has not been worked into the contractual agreements, or other waivers have not been signed because the architects and engineers don't want to legally expose themselves to misinterpretation of information any more than they already do.

Communication and Collaboration Methods

Depending on the type and schedule of the project, information is often not shared with the construction team until the 100 percent construction documentation (CD) phase. Drawings are distributed from the architect or local print shop, at which point the contractor then takes either the sheet drawings or the digital PDF and CAD files and performs a *takeoff*. The takeoff process, in the case of the sheet drawings, is done using document tracing software, manual takeoff, or on a digitizer (Figure 1.5).

Figure 1.5 Using a digitizer to complete an estimate takeoff

A digitizer helps an estimator trace documents and quantify items such as walls, floors, and ceilings and counts the number of doors, fixtures, and equipment, while also flipping through corresponding drawings to see whether the drawings paint a clear picture of the design intent as communicated by the architects and engineers. This process is disconnected in that it relies entirely on the ability of the estimator to correctly interpret drawings that are assumed to be accurate. The problem with the digitizer method of takeoff is that the level of interpretation left to the person doing it is great; often, a significant amount of misinterpreted data is input into the estimate. The other resource that is being consumed in great quantity here is time. Especially on larger, more sophisticated projects where it is critical that data is correct, this method takes a large amount of time and effort.

This method of delivery often requires numerous clarifications, which are ultimately the sole means of communicating with the design team aside from site or prebid meetings. Although this method tracks the questions being asked and the answers being issued, it is usually too cumbersome to navigate in a relatively short period of time to provide an effective means of project communication. Often, the big issues are addressed, and smaller issues that aren't understood are interpreted and assigned a contingency to be resolved later.

Type of Documents

The typical documentation for design-bid-build is printed sheet drawings and specifications. The practice of providing the contractor with a PDF, CAD, or image file has become more commonplace, and estimating programs such as On-Screen Takeoff, SOFTakeoff, and Bluebeam can speed up this tracing process. Contractually, however, the design team is not typically responsible for sharing digital files and often doesn't—to limit any involuntary editing or further possibility for misinterpretation of information. Although the estimating software saves time, this system relies on the estimator to quantify accurately all the building components in a set of drawings and assign prices and estimates to the labor, equipment, and materials associated with that construction, using the architects' and engineers' design drawings for accuracy. Although this is the typical responsibility of the estimator, the issue of time continues to come into play. The reasons for this are that in a design-bid-build delivery, the architects and engineers have been working and coordinating drawings for much longer than the contractor, who typically has only weeks to fully understand the site, systems, construction, and reasoning behind the design before assigning a cost to the project. Although much of the project's estimate basis might be determined by a square-foot cost and subcontractor input, there is a large margin for error because of the lack of time to fully understand the project and all of its nuances.

Clarification of Information

Clarifications are formally issued and addressed in addendums that include supplemental drawings from the architect or engineer, specification clarifications, and other directives. These clarifications are then distributed to all contractors bidding on the project via email, fax, project website, distribution list or public notice. Additionally, these clarifications must be tracked and audited as to when they were issued and that all bidding parties received the clarifications. Again, in this type of delivery, the contractor and subcontractor have no input in the actual design and documentation process, with the exception of clarification and supplemental drawing information and are ultimately responsible for checking with the design team, owner or owner's representative to verify receiving any further updates.

Once the contractor has created an estimate for the project, the contractor's bid is based on the information provided by the design team and often carries a contingency to allow for information missing from the documents that is later resolved in the field. In this type of project, CAD drawing information is to be built as drawn by the architect and engineer. Because of a lack of flexibility, this process leaves little room for adjustment during construction, which may lead to an adversarial relationship between the designers and the contractors performing the work. The reason for change orders often involves the contractor's requirement to construct something that may be considered unbuildable as drawn. Another reason is that the contractor might have means of building something more efficiently than what was drawn by using new technologies, past experiences or new tools that the design team wasn't aware of when they were creating the documents. While this change may equal a price deduct, a change order will need to be issued to address this change. This lack of flexibility may equal additional costs when local jurisdictions having authority (JHA), standards and governing building codes require a certain type of construction that the contractor may have been aware of and that the design team might not have been. As the drawings and specifications in the design-bid-build method are the sole means of communicating exactly what is to be built, when situations arise that weren't thought about in the design and documentation phase, the contractor issues a change order. This is because every item that wasn't assigned a cost in the initial construction documents is considered "in addition" to original project scope and leads to extra costs. That said, incorrect documentation and lack of collaboration equal more costs, change orders, and inefficiencies in this delivery method.

Project Closeout

At the end of a design-bid-build project, the CAD files, shop drawings, specifications, RFIs, and change orders are compiled into a binder and submitted with a operations and maintenance (O&M) manual. These documents are submitted to the building owner after the final walkthrough and the completion of construction work under contract. This usually marks the end of the responsibilities for the contractor.

Often the O&M information turned over to the facility manager is an inadequately organized mixture of disconnected information. The facility manager is then tasked with inputting additional information or layers of relevant information over this compilation of disconnected data. This information includes tasks such as work orders and maintenance requests, move orders, associate locations, telephone extensions, equipment warranty information, emergency exit strategies, and any site-specific facility information such as laboratory clean rooms, hospital head walls, sensitive government data, and so on. The CAD files delivered to an owner are usually unreferenced or part of the architecture firm's legacy information, which might involve customized plug-ins, applications, and routines that are unable to be opened by the new facility manager.

This delivery method can drive a wedge between architect and contractor, especially if the construction documents aren't precise enough to cover work included in the contractor's contingency. This delivery is a perfect example of an old way of thinking, using a rigid system of information management and sharing, where the main focus is to avoid litigation. The architect is responsible for documenting all work to be completed on the project. Total documentation, especially through CAD, is nearly impossible in any project, and too often time is spent drafting details and views to prevent misinterpretation, as opposed to staying focused on the design and the owner's desires and requirements for the project. BIM in this model can be used little, aside from efficiencies realized by the engineers and architects using it to better coordinate their design documents and some use by the contractor for quantity extraction. Additionally this method promotes the separate creation of a construction BIM used in the field, which is developed by the general contractor separate from any construction documents and holding no design professional's sign or stamp. This in turn creates additional liability, which will be discussed in detail later.

Design-Build

The Design-Build Institute of America (DBIA) says this about the design-build method of delivery:

> The design-build form of project delivery is a system of contracting whereby one entity performs both architectural/engineering and construction under one single contract. Under this arrangement, the design-builder warrants to the contracting agency that it will produce design documents that are complete and free from error (design-builder takes the risk). The selection process under design-build contracting can be in the form of a negotiated process involving one or more contracts, or a competitive process based on some combination of price, duration, and proposer qualifications. Portions of the overall design or construction work can be performed by the design-build entity or subcontracted out to other companies that may or may not be part of the design-build team.

—*An Introduction to Design-Build*
(Design-Build Institute of America, 1994)

Many envision design-build as the BIM solution. Design-build delivery *is* much more integrated than design-bid-build and with the introduction of a design assist agreement, can create a strong foundation for collaborative practice. The design-assist agreement dovetails into a typical design-build contract and allows for the contracting team to have early involvement in a project, with a concession for the potential to recapture the fee when the design portion ends, if not selected for the project.

Although the DBIA holds no specific BIM contracts currently, it does strongly promote the early formation and collaboration of project teams. This might change as more owners, and specifically those who most often utilize the design-build form of agreement, demand BIM. Ultimately, the framework of design-build is structured to facilitate the use of BIM. However, some of the typical project deliverable timelines will need to be shifted to facilitate creating BIM documentation as opposed to CAD documentation to facilitate the new resources and tools available to construction managers to deliver a better project.

Preconstruction

In design-build delivery, the contractor or architect is contracted as a single entity known as the *design-builder* or *design-build contractor*. The purpose of this type of contract is to increase accountability and have a single source of project delivery. In this type of system (Figure 1.6), the design-builder is responsible for streamlining the process by combining the design, permit, and construction tasks. If the lead is the architect, the contract is for a "design-led design-build" project. If the lead is the contractor, the contract is for a "contractor-led design-build" project. In either case, both parties are under agreement to design and construct the owner's building in budget and on time.

The rising popularity of design-build shows it to be one of the more effective ways of delivering a project. However, there can be inaccuracies and ambiguities in this process because the construction can happen in parallel with the completion of the design documentation. The process is weak in design review because the design is still being completed as the project is being constructed. Quality control tasks associated with the design team become secondary, because the primary goal becomes completion of the project under a contractor-led agreement. The quality of design produced by the architect and design team can suffer as well, because the contractor's responsibility of coordinating trades and schedules on a working construction site becomes the driving factor for the project, not aesthetics.

Conversely, in an architect-led design-build project, there is the potential for the focus to become the aesthetic and design elements of a project instead of the project schedule or other construction-related tasks. A fundamental issue with a design-build project is that ultimately one project team member has seniority over the other by default. The fact is that whether it is the contractor or the architect, by choosing one or the other, the project team is not all on a level playing field, which can ultimately lead to project complications later. Design-build's efficiencies are in overlapping the design phase with the construction phase to shorten schedule and reduce project costs. To efficiently use the design-build delivery, you need a balance among the team members, built upon a schedule that enables the use of BIM processes. Remember that BIM is not necessarily a fast-track means of delivering a project. Rather, it is a technology that allows for more coordination before the project is constructed due to streamlined documentation processes.

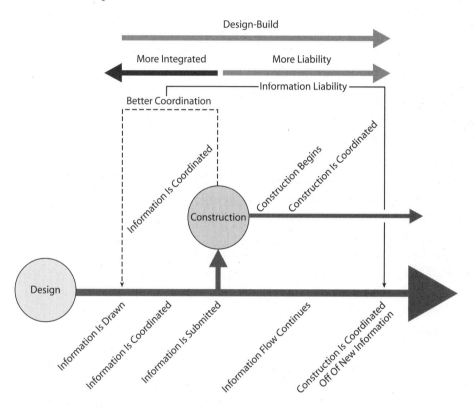

Figure 1.6 Design-build information flow

Communication and Collaboration Methods

Information flow in this type of project delivery begins with the initial design produced by the architect, as presented to the owner for review. This design is then used by the contractor to begin putting together an estimate and schedule for the project. As many architects and contractors know, the first design is rarely the one chosen and built. So while the contractor is assigning a cost to the first design proposal, the design is already outdated and incorrect; the architect is now revising the design per the owner's design changes and contractor input. This continues throughout the project process, because the architect is constantly trying to stay ahead of the contractor and the contractor is trying to catch up to the architect's design drawings.

The construction documentation phase of a design-build process often begins with 50 percent of the construction documents going to local code authorities to secure a permit. Construction planning and site development begin at this point in anticipation of 100 percent of the drawings being finalized. The mechanical, electrical, and plumbing engineers' drawings, as well as the specifications, are submitted in tandem with these documents. Engineers in a design-build contract are contracted directly with either the architect or the contractor and typically engage the design at the design development level of the project and in some special cases during schematic design. The engineers then begin to create their layouts, complete calculations, and size equipment based on the architect's design. The contractor then begins to assign costs to the engineers' layouts as well as the architect's drawings, while simultaneously beginning construction on the project. The construction of a building while design documents are being completed is unique to a design-build process. For example, allowing for the construction of certain packages of work, such as concrete or steel, requires careful coordination with the designers and engineers to make sure that as their designs are being completed they don't alter or interfere with work already being done. Although this is an opportunity for BIM to shine in this type of delivery, it is also a challenge to constantly update the composite BIM with new information from the architects and engineers.

Types of Documents

In a typical design-build project, documents include printed construction documents and specifications, CAD files, and PDF files. In a design-build agreement, CAD and PDF files are readily shared, because the team is responsible for building the project together; as one benefits, so does the other. For this reason, the formality and rigor of document sharing are reduced when compared to the design-bid-build means of delivery. Sometimes, because of the sensitive nature of a firm's legacy data, the architect or engineer will require a media release or a nondisclosure agreement to be signed by the project team. This is a means of protecting the firm's database of information from being shared with competition, either intentionally or unintentionally. Many firms deal with company-specific digital information by printing the drawing information into a hard copy or PDF and then deleting the native files from the shared documents in order to avoid any issues.

The same agreement might be required if BIM is integrated into this type of delivery method and specifically if the general contractor has an in-house design department but is working separately from that department on a project. A BIM type of project documentation can be planned for and coordinated in a design-build process and should be introduced in the initial contract negotiation meetings. Chapter 2 discusses how to build and integrate an information exchange (IE) responsibility plan and a model coordination plan. Both of these, or similar documentation, should be required in a design-build project if the intention is to use BIM in any fashion on a project.

Clarification of Information

Changes in this process are addressed with cost updates. The preliminary contracts usually provide for design alterations and changes throughout the design and construction documentation process. Typically a point of no return takes place in the project, after a final budget has been issued, when design alterations stop and further design changes result in additional project cost or change orders in the field.

Conceptually, design-build aims to limit the exposure of uncoordinated items and, through collaboration, increase the viability and accuracy of the project. Yet this process also relies heavily on the integrity of the contractor to deliver the project within budget and schedule, which may be difficult because the quality and interpretation of design documents leaves room for misinterpretation and assumption. Although not all design-build projects are fast-track or require additional construction and design coordination, many times the project is similar in timeline and schedule to a design-bid-build delivery, with the major exception that the project team is integrated. A rising perception within the industry is that—just like cheaper isn't better in a design-bid-build project—faster isn't better in a design-build project. In actuality, the more coordination and clarification that can be accomplished before a shovel ever touches dirt, the more potential issues can be avoided later.

Many times in design-build the engineering team provides a performance specification. It is then left to the subcontractor to design and build a system that meets these requirements. Many subcontractors are familiar with this and go about designing and issuing shop drawings for engineering approval. Yet some companies have seen a unique opportunity. Because they ultimately design the mechanical, electrical, or plumbing system and build it, there has been a rise in companies integrating engineering in-house and offering both services. By streamlining internal processes between the engineer and the fabrication shop, many of these companies are becoming more popular, specifically among more integrated teams, because of the coordination they can offer.

Project Closeout

At the completion of a project using design-build, the O&M manual is issued, along with hard copies of the building drawings, shop drawings, field changes, specifications, change orders, and punch lists. This information is not in a connected format and often is a hybrid of paper and digital documents, just as in other delivery methods. It then becomes the responsibility of the facility manager to correlate this information into usable documentation.

BIM in design-build presents a unique opportunity by allowing facility managers to define early on what they expect to see as a deliverable at closeout, not only the type of documentation but also the level of detail within the documentation. The buzzword of *digital O&M manuals* pertains to the concept of embedding within BIM components relevant and specific information. Information such as cut sheets, photos, shop drawings, pictures, and URLs can potentially be inserted or linked to model components (see Chapter 7). Combined with a more integrated means of delivery, design-build offers unique opportunities as a delivery method for BIM projects.

Design-build is the father of a true BIM process. It introduced the idea that a project team that collaboratively seeks to complete a project can realize efficiencies and profits. Design-build delivery continues to be a good starting point for those interested in beginning integration one step at a time, as well as a means of building a BIM process through hybrid documentation.

Note: It is always important to consult with legal counsel prior to engaging in or using altered or untested contracts, agreements, and plans. Although the examples in this book aim to further define what tasks are important in a BIM process, you should always review all documentation with your legal counsel.

CM-at-Risk

CM-at-risk entails a commitment by the construction manager to deliver the project within a guaranteed maximum price (GMP). The construction manager acts as consultant to the owner in the development and design phases (often referred to as *preconstruction services*) but as the equivalent of a general contractor during the construction phase. When a construction manager is bound to a GMP, the fundamental character of the relationship is changed. Not only does the construction manager act in the owner's interest, but the construction manager must manage and control construction costs to not exceed the GMP, which would be a reduction in fee and as a result a loss in profit.

> *One of the most important aspects of the survey results is the changing attitudes concerning construction delivery methods. Quasi-public and government organizations predominantly use the design-bid-build method, but clearly, many have tried other methods and most would consider either CM-at-risk or design-build to be the best-value alternatives. Changing the delivery methods used, in the case of these organizations, will often require changing laws and politics, but that is happening, too, because the public is best served when it gets the best value for its tax dollars. Privately held and public companies continue to try a variety of delivery methods...but CM-at-risk will likely become the more dominant delivery method for this group as long as the experience is positive.*

—FMI/CMAA Sixth Annual Survey of Owners (FMI, 2005)

CM-at-risk delivery methods can be customized to a BIM process. CM-at-risk as a BIM process has two key ingredients. The first is that there is a belief in the industry that a more integrated process equals a more profitable one (Figure 1.7). The second ingredient is a perceived value in leveraging BIM technology with the team.

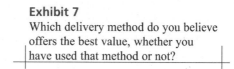

Exhibit 7
Which delivery method do you believe offers the best value, whether you have used that method or not?

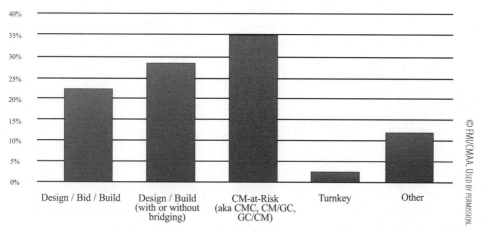

Figure 1.7 Perceived delivery method value

Preconstruction

The flow of information in a CM-at-risk contract can provide an integrated service. By *integration*, I mean the ability for the contractor and subcontractors to be involved with the project very early on and have input into the design and documentation of a building. The CM-at-risk model puts the risk for delivering the project at the proposed GMP on the contractor's shoulders and thereby gives the contractor a stake in the development of the project. What is valuable from this type of delivery is having the contractor sitting at the same table as the design team.

During preconstruction, the CM's involvement is critical to the success of this type of project delivery. The contractor can continually inform the design team of cost based on the current documentation. Using a design-to-budget approach, the contractor removes the value engineering period associated with project delivery methods that typically come in over budget. *Value engineering* is the belief that by allowing time for the design team to redesign a project to attempt to reduce cost, the changes made will save the project money. This concept is flawed; as a process, it indicates only that proper estimating procedures were not in place prior to the design being completed.

Although this process is prevalent in design-bid-build and even some design-build projects, the CM-at-risk delivery method somewhat mitigates this issue because the contractor is intimately involved with the estimating process of the project, because they are required to deliver a GMP based on the completed design.

Communication and Collaboration

The contractor in this method is responsible for delivering the project within the budget. The architect who contractually may be under the contractor or the owner, while attempting not to exceed the project budget, may deliver a project that programmatically and aesthetically pleases the owner. The risk to the owner in CM-at-risk is the contractor taking too much control of a project, especially when the contractor enters the realm of design and owner program management. If the project is difficult and unique in design, a construction manager may become concerned with the cost and difficulty of the project. By providing estimates and updating material costs, the CM enables the owner to make a decision based on the aesthetic or the cost and move forward without damaging the project timeline. The real issue for this type of project delivery is making sure there is an involved owner in the process to see that both the design and the cost requirements are being met.

Ultimately, the contractor is responsible for delivering the project on time and within budget. However, that shouldn't come at the expense of good design and project balance. By seeking project balance and collaboration in the flow of information and management of the project, CM-at-risk can effectively utilize team integration and BIM technologies. With proper up-front coordination and planning, the CM-at-risk delivery method is an effective means of bringing all team members to the table and sharing responsibilities equally among them.

Type of Documents

The typical documentation for a CM-at-risk delivery is printed contract documents. Again, because of team integration and depending on team agreements, PDFs and CAD files may be made readily available. As the design develops, the need to continually update the estimate may affect the means of transferring data. In some cases, a project FTP is established, or a means of drawing distribution is established through either a print shop or plan distributor. In other cases, the drawings may be simply emailed through a point person who tracks and archives the files that were sent for future reference.

In this delivery method, the need for agility and rapid transfer of data is primed for BIM. Using a composite model, multiple design changes can be housed in a single model and can easily be imported to replace antiquated data, which can then be archived. BIM holds an enormous advantage over CAD in this type of delivery. The three-dimensional construction of a facility inherently holds quantitative information that may be used early in the process to establish a preliminary estimate and coordination. In addition, the cumbersome management of multiple singular drawings or CAD files associated with each profession for each update can be overwhelming, whereas a BIM is a single file to update that contains all the necessary information relative to that profession.

Clarification of Information

The process of clarifying information with a CM-at-risk delivery is integrated and project focused. Clarification during preconstruction involves direct interaction and input from the general contractor and even subcontractors. The contractor is able to clarify a number of issues, including budget, estimate breakdown, trade coordination, and constructability. By providing a GMP for the project, the contractor has a vested interest in providing the design team with as accurate of information as possible. Likewise, the architect and engineers have an obligation to the contractor to provide as much information as possible along the process of design development to further refine the scope, budget, and schedule of the project.

During construction, the contractor is typically very pliable and, instead of taking an adversarial approach to issues that arise, takes on a mediator role. This is because profitability is directly tied to the contractor's performance and project coordination. While bidding to subcontractors, if required, the contractor and design team have it in their best interest to give the subcontractors as much information as possible about the project to improve the accuracy of the estimate and to reduce any large contingencies. Although many of the issues should be resolved prior to construction because of integration and team involvement, there exists the potential for a general contractor to receive an additional bid for a scope of work if they believe the estimate to be too high.

BIM fits well into the CM-at-risk method of delivery. The BIM tools available allow for the ability to test and coordinate prior to construction, thus limiting the need for clarifications. Yet if clarifications are needed, BIM provides the ability to quickly find answers, which is critical in a CM-at-risk project where large amounts of data are being frequently moved.

Project Closeout

At project closeout, the owner receives information just as in the other methodologies. The facility manager is again responsible for coordinating all the documentation, information, and correspondence as part of the job. In some cases, the facility manager is brought to the table early and defines what the expectations of project closeout deliverables might be. This early interaction should also be written into the project contracts as required deliverables for the project, because otherwise the facility manager is left with the same jumble of information as with other methods.

Often facility managers are hired after the completion of the facility and are not as familiar with the facility as the project team and owners. Therefore, the flow of information and project experience are disconnected. As a best practice, construction managers should ask what type of deliverable is expected at project closeout for a number of reasons—first, to define the cost and resources needed to deliver the documents, and second, by being prepared and asking, the construction manager averts dissatisfaction from the client and provides the new facility manager with the requested data.

A New Concept of Delivery

Integrated project delivery (IPD) is a new form of project delivery that has gained popularity as an integrated solution. Although many firms have practiced integration, this new definition of project delivery and contract language aims to take it to a new level.

> *Integrated Project Delivery (IPD) is a project delivery approach that integrates people, systems, business structures and practices into a process that collaboratively harnesses the talents and insights of all participants to reduce waste and optimize efficiency through all phases of design, fabrication and construction.*
>
> —"A Working Definition—Integrated Project Delivery"
> (AIA California Council, 2007)

Preconstruction

IPD calls for a complete integration of teams from the onset of a project, allowing the team as a whole to become a collaborative group that focuses on leveraging the latest technology to foster flexibility and successful project outcomes. This delivery method has really started to set the stage for a truly collaborative process. Although varying degrees of a BIM process can be used in virtually every delivery method, this method allows for a greater degree of potency in the process and promotes project balance through the required use of BIM. George Elvin makes an excellent case for integration and spells out how it's critical to the success of the industry.

> *Pioneers in integrated practice are finding they can amplify their fees, expand their services, and build long-term relationships with their clients by working in a highly collaborative relationships with all project stakeholders throughout the complete lifecycle of the buildings they create.*
>
> —George Elvin, *Integrated Practice in Architecture* (Wiley, 2007)

Communication and Collaboration Methods

By integrating BIM technology and using new delivery methods that focus not only on the successful delivery of the project but also on project balance, rewards are achieved in the form of profit, professional relationships, reputation, and money. A fundamental flaw in all the previous delivery methods is value added vs. project cost. In most scenarios, the project team is reimbursed as a percentage of the project cost. This quantifies in some way the scope of the work to be performed by the project team. The flaw is when a member of the project team or the project team as a whole improves collaboration and creates value or savings for the project. This results in the following:

- There is no incentive for the AEC team to create any additional value, because there is no additional compensation for the additional resources required to further collaborate.

- If the professional's fee is based on a percentage of the project, the fee may be reduced for the professional because of significant project savings.

IPD promotes the concept that by sharing the risk and reward of a project through target project goals, that compensation may increase or decrease depending on results. As an example, the team, including the owner, develops a goal for the entire project budget. If the project comes in under budget, then additional fees are distributed to the team; if the project comes in over budget, fees are reduced. By holding the others accountable, IPD fosters a great degree of communication and promotes intense collaboration among the project team, because it can result in additional profits.

This delivery method involves the entire project team from very early on in a project and consists of project goals, which are shared and incentivized throughout the team. By using the knowledge of *all* parties, including subcontractors, consultants, and local governing bodies, IPD aims to eliminate issues in the field that result in significant cost overruns later in the project. Through increased accountability and promoting teamwork, IPD is a model for new process teamed with new technology.

NELSON-ATKINS MUSEUM ADDITION, KANSAS CITY, MISSOURI, CONSTRUCTED UTILIZING BIM TECHNOLOGY. PHOTO BY BRAD HARDIN.

Types of Documents

IPD is unique in that it is driven by BIM technology. IPD relies on BIM not only to be more collaborative and integrated but also to be a quick and efficient means of developing a project. With BIM, a change to one element equals a change everywhere;

this means that the technology is limber enough for a design to be developed, tested, altered, and updated during preconstruction to eliminate coordination issues later.

Documentation in an IPD process is a combination of individual profession-focused models, such as the architectural and engineering models, and the composite BIM documentation. This BIM documentation can be used for estimate revisions, constructability reviews, clash detection, site coordination, and a host of other coordination responsibilities. Because many changes can be represented in one model file, the number of information transfers is reduced, but the information is able to be tested and coordinated more quickly than in CAD.

Clarification of Information

Information flow in an IPD process (Figure 1.8) continually informs the team and allows the project stakeholders to have a say in the project and make informed decisions as a whole. The advantage of this type of information management is that the biggest focus of the project now becomes using and sharing information. It is no longer the litigious arena that architects and contractors have played in for decades but rather a new platform that effectively challenges the knowledge base and experience of a project team by making the focus understanding and early issue resolution, as opposed to profession-focused concerns.

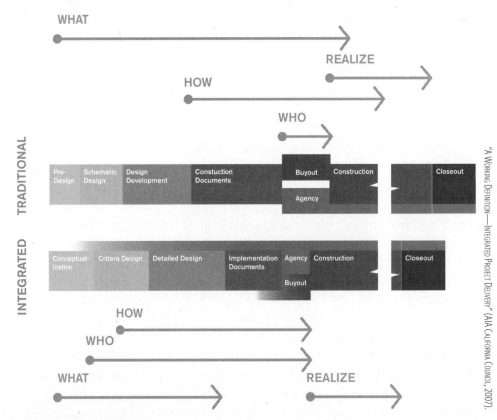

"A WORKING DEFINITION—INTEGRATED PROJECT DELIVERY" (AIA CALIFORNIA COUNCIL, 2007).

Figure 1.8 Integrated project delivery

Project Closeout

Integrated project delivery provides the stage for all the team members to perform at their best. IPD is unique in its ability for the facility manager to be involved with the construction and design of the building and ultimately to use BIM as a tool to better maintain the facility. Information sharing and data management techniques are refined to make an integrated project successful. A completed BIM at project closeout avoids the typical disconnected data and provides the facility manager with a much more useable tool than CAD. Other documentation still must be compiled either digitally, embedded in the model itself (see Chapter 7), or embedded in an O&M manual.

A New Concept of Process

For decades, the way data was transferred in a construction process was linear, with one task being followed by another. This old process of drawing collaboration involved a large number of isolated files that needed to be continually updated and constantly managed. In a typical CAD process, information was created in isolated CAD (or other) files and then digitally inserted over the top of the floor plans, sections and other separate views. For example, the architect would reference a single mechanical floor plan file of a single floor and use the single structural engineers' CAD drawings for a single floor as a background to test for conflict interference between the two. This process was time intensive, not automated, and cumbersome to review in a project with frequent updates.

In the old system of information management, documentation was updated by deleting or archiving CAD files and then inserting the updated drawings over the top of the new files. A disconnection of information often happens here. For example, say the architect has to redesign a wing of a hospital or the owner decides to move 30 percent of a building's square footage into the next phase of their building plan to stay within the budget. A lot of work is associated with these kinds of changes. The architect now needs to send out updated plans and other drawings to all of the MEP, structural, and civil engineers. Subsequently, these engineers must redraw any items that directly influence their trades. As drawings are completed, they are redistributed back to the architect. Now there is more coordination, not only between the architect and the engineering team but also between the plumbing engineer, who needs to know where the new plumbing mains are to be routed from the street, and the electrical engineer, who needs to know, because of the lighting changes, if they are now hitting the bottom of the redesigned ductwork in the system. There are lots of possibilities for coordination to be missed in this process, especially with tradespeople, who might not be specifically thinking about the other trades' responsibilities within an area. This coordination of information silos is the first big area for concern of the old system of information management.

The second area of concern is the time it takes to reconstruct and update information. How long will it take the architect to now overlay the MEP drawings and go through the new drawings with a fine-tooth comb to find any conflicts as well as get their own drawing tasks accomplished? It is often a monumental task that only gets worse the more complex the building and more demanding the client's needs, budget, and occupant requests. In addition, this cuts into the profits of the design team, whose fee was based on a specific scope of work.

Where Does a Contractor Fit in a BIM Process?

The contractor should be considered a member of the design team, allowed not only to manage construction but also to help manage the information that is being communicated in order to build the facility. I am not suggesting that contractors take the place of architects or engineers but rather that they be considered just as valuable a member of the design team as the architects and engineers.

The main purpose of construction documents is to communicate the design intent to the builders from the architect. In the past this process has been linear; in this new type of process, the information is sequential: one event defines the event that follows, and then that defines another event and then another. The model is a way to virtually construct, test, change, and communicate design intent in a way that wasn't available to the construction team before. Thus, BIM informs designers so they can make better decisions about their designs. Conversely, it allows the contractor to determine the means and methods the contractor anticipates using to build their visions as well as provide a contractor's perspective on the design. The reward for the contractor being engaged in the construction documentation or the implementation of the documentation phase is that it provides everyone on the team with insight into the actual construction of the project. In a BIM project, while the drawing/modeling time is front-loaded, the size of the staff and the amount of time required to model the project are both reduced. The big idea in a BIM process is not only the ability to store information within the model but also the ability to communicate better. For example, using a 3D rendering during design to coordinate a piping layout makes for more informed decisions using the model.

If you start to think of the model as a virtual construction of the finished product, the question becomes, how many ways can this information be used to help the design team from a contractor's perspective? This book gives specific examples of integrating BIM in estimating, site coordination, construction coordination, enhanced schedule visualization, phasing, trade coordination, clash detection, sustainability issues, improved in-field communication, 3D shop drawings, and facility management models that go beyond making the case for BIM, but rather define a whole new level of virtualization.

Using BIM also inherently improves communication. Where before we could see lines on a light table or overlaid on a computer screen, now we can see in a 3D view. Before BIM, projects consisted of coordinated plan, elevation, and section information to understand the building; now we can create sections, elevations, and perspectives quickly in real time because they are all different slices, views, and projections of the virtually constructed model.

Integrating project teams before project delivery also promotes team buy-in. As opposed to the adversarial relationships that could potentially develop between team members, a truly integrated process limits potential litigation in a BIM-focused process. The contractor understands the design intent more fully than before, as well as feeling that they have used their knowledge of construction to further inform the design team about decisions that involve actually building the design. Another great aspect about BIM is that it reduces contingencies. According to Michael Kenig's article "Beyond BIM: Spending Money to Save Money" (*Constructor*, September/October 2007), on average 1 percent of a project's cost is spent on resolving non-owner-initiated changes. He goes on to say that with the use of technology-enabled coordination on a project, BIM can reduce contingencies and increase the return on investment in BIM by three to four times.

Integration using technology permits the contractor and subcontractors to be familiar with what is to be built. In addition, the contractor can clarify the scope and budget earlier in this process, prior to bidding or construction. Ultimately, a contractor's knowledge, experience, and input is more valued and utilized in a technology-applied integrated process than in any other means of delivering a project.

Is BIM a Fad, or Is It Here to Stay?

The people asking whether BIM is a fad are not asking whether BIM is the way the future of design and construction is headed but are rather asking, how much will I have to change? and how much will I have to learn? As new technology continues to emerge, it's good to ask questions about software and applications with regard to the best way to implement these new tools and whether they are right for your use. Don't avoid BIM technology because you are not comfortable with change. We live in a world that is constantly changing and adapting. New technologies are a direct result of that constant change, and many companies see BIM as an opportunity to make themselves more profitable and implement tools that owners, new employees, and contractors enjoy using. The fact is that the industry is not racing back to drawing lines in CAD or even creating hand-drawn line drawings, as beautiful as they are.

The reality is that the productivity loss in the construction industry because of inadequate coordination equates to about $60 billion a year, according to Michael P. Gallaher, Alan C. O'Connor, John L. Dettbarn, Jr., and Linda T. Gilday in *Cost Analysis of inadequate Interoperability in the US Capital Facilities Industry* (NIST, 2004). This, combined with the exponential growth of BIM in the job market (Figure 1.9), begins to paint a clearer picture of the need for both technology and people who can better coordinate construction projects.

Figure 1.9 Job listings calling for selected skills

Technology and time are two of the biggest threats to the successful completion of a project. Rapid developments in technology make for dependence on everything from cell phones, emails, software and equipment to better coordinate projects. If these tools don't work, it makes it very difficult. Secondly, time is critical to any construction project. A good construction manager knows time is important and uses all available tools to ensure a project's successful delivery. But, technology and time are also the biggest potential advantages. Specifically, the use of technology over time develops, creates, and refines tools for its users. Technology rarely goes backward. Technology continues to move forward at a blistering pace. The use of BIM has been no different. Within the past decade their have been huge strides in BIM technology, which have equated to a significant rise in users. These industry professionals have "spoken with their pocketbook" as to the value of BIM. Specifically in the construction industry, many companies see the investment more than worthwhile and use it to market themselves against competitors.

Although many remain skeptical about BIM and BIM-related technologies, you need to ask whether you are being the most productive and profitable that you can possibly imagine. If the answer is no, then it's apparent something needs to change in order to make it that way. To that end, I strongly suggest investigating BIM.

Ten Steps for Successfully Implementing BIM

What does it take to implement BIM? When you start down this path, you have to ask some questions. What are you trying to achieve with BIM? What elements define this company? And what steps are necessary to begin implementing BIM software and processes?

To begin, develop a simple statement about how BIM aligns with the goals of the company, how it can be used in the future at the company, and how it might make your company more successful. This brief statement should define the organization's stance on new technology. This will become vital information later, in the implementation phase, when the pieces of software that have been identified by the organization to implement might go beyond BIM. Additionally, ownership needs to be involved in this initial discussion of strategy, because they will have to decide on investments in software, hardware, and staff.

Ten steps are critical to the successful implementation of BIM in any organization, outlined in the following sections.

Step 1: Identify a BIM Manager

When a construction company embarks on constructing a structure, the organization staffs a project manager to direct and organize the project. This is the same in virtual construction. Similar to a construction manager, a BIM manager must manage and facilitate all the processes necessary to create and manage BIM. This involves coordinating all the information from architects, consulting engineers, and subcontractors. The BIM manager also coordinates project reference points and develops a schedule that identifies when tasks such as clash detection and model updating need to take place. Overall, the BIM manager needs to have old skills, new skills, and, most importantly, an open mind and ability to solve problems. In his article "The New 'Must Have'—The BIM Manager," Dominic Gallello outlines the responsibilities of a BIM manager as follows:

- *Understanding project workflows (schematic design, design development, construction documentation phases) and project management.*

- *Understanding different needs of the delivery team (architects, engineers, estimators and contractors). The BIM Manager works much earlier with the entire project team in setting up the project structure and data exchange formats.*

- *Technical knowledge of the BIM application used, related systems and network infrastructure, and awareness of new technologies.*

- *Communication and training skills (verbal and written).*

- *Strong teaching and coaching skills to bring new team members up to speed.*

- *Ability to communicate the benefits of BIM firm-wide, including the "personal win" at each level in the organization.*

- *Objective decision-making in times of crisis.*

- *Flexibility and mobility. Large multinational firms with multiple offices worldwide often require BIM Managers to help the implementation of new company standards throughout the whole company. In addition to a desire to see the world, being sensitive to cultural nuances will be a great asset.*

—HTTP://WWW.AECBYTES.COM/VIEWPOINT/2008/ISSUE _ 34.HTML

Many companies choose to start the process with a single professional internally—someone who has good management skills and who has a background in technology. This is wise, because the best person for this job needs to have an intimate understanding of the day-to-day functions within the company. If this resource must be a new hire, then it is critical to choose an individual who is highly competent in organizational and communication skills, has a background in BIM technology, and can be trained in different pieces of software and to manage multiple tasks.

This BIM manager becomes a key player in the next nine steps of implementation. Before selecting the BIM manager, consider the manager's involvement on other projects, because the implementation process is time-consuming and will become the sole initiative of this individual. This person needs to be able to understand the functions of the software and how it will work with the company's operations. In addition, it will be the responsibility of this BIM manager to spearhead the process of integrating BIM into the company. The goal of the BIM manager is to identify what will work best for the company and make recommendations to the leadership about what is valuable and what might not be the best fit or might need to be further developed.

Step 2: Develop an Estimate of Cost and Time to Implement and Use BIM Software

The next step is to put together a software and hardware acquisition plan. This plan should include the cost of the software, the hardware, and any additional staff needed. The goal of this plan is to give management an idea of the scale of the investment needed. It should include yearly subscription costs, support costs for at least the first year of using the software, and any other costs associated with using the software. Potential hardware costs include additional RAM, disk space, servers, or network connections that are required. The software vendor can generally furnish this information.

The following is an example of a line-item estimate for one user to begin using BIM with an extremely robust set of tools:

Equipment	Cost	Time Period
Dell Precision M90, with additional memory and enhanced graphics card	$2,400	
Microsoft Office tools or equivalent	$300	
Architectural CAD/BIM modeling software	$3,200	First year
Structural or energy analysis software	$1,000	
Estimating software	$7.200	
BIM model compiling software, such as Navis	$9,300	
32″ HDTV LCD monitor (optional)	$1,200	
Video projector (optional)	$600	
FTP site service provider annual charge (optional)	$1,900	
All software's annual service charges (subscriptions)	$1,200	After first year
CAD training charges	$6,000	

Equipment	Cost	Time Period
IT cost for initial setup	$2,200	
Dedicated large-format plotter/printer and service charges	$2,800	
Annual salary of staff	$70,000	Annually
Cost of attending industry events such as seminars, trade shows, and peer group events	$4,000	
Grand total	**$112,100**	

Although this example shows a full-blown, robust BIM machine, training and software, you should keep it in perspective. BIM is an investment and requires a significant cost; on the other hand, the potential savings and return on investment far outweigh the costs of hardware and software, and can be purchased over time.

Because many of the pieces of software require additional horsepower to make the software function correctly, this can make for a significant investment by the firm. Further development of the plan should include a description of each piece of proposed software, a rationale for its use, the cost, and estimates for the time to implement it and train personnel on its use (see Figure 1.10).

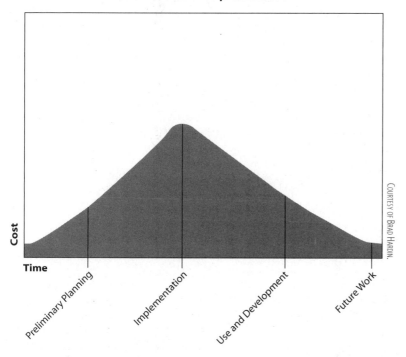

Figure 1.10 Time vs. cost of implementation

Implementing a BIM solution is an endeavor in itself; to make the overall transition easier, a firm should not try to acquire and train people on multiple pieces of software at the same time. Identify specific pieces of software in the estimate that show the initial investment and time, and show what software is to be acquired later in the integration plan.

The goal of the acquisition plan is to give management a clear understanding of the total cost to implement the proposed solution and to secure ownership buy-in. Ownership may begin a conversation about which software products can get the firm to walk in BIM before everyone has to start running in it. The BIM manager should rely on the experience and guidance of management and senior staff to help develop a plan that everyone can support.

Step 3: Develop an Integration Plan

The implementation plan consists of a software acquisition plan, a training schedule, a hardware update schedule, and a narrative explaining the company's shift into this new technology. Additionally, the implementation plan explains how the BIM strategy will be rolled out across the company. This plan will take time to build, so account for this.

For larger organizations with multiple offices across multiple states and a large employee base, it's best to start with a single office that can become the hub for the system. You won't gain anything by attempting to implement BIM at two or more locations at the same time. The BIM manager needs to research and interview the different departments to find out what software is currently being used as well as what processes are in place in the organization. Often it is helpful to list the software and the departments in a spreadsheet to analyze what existing software is BIM compatible.

For smaller companies, take inventory of what is currently being used, and then develop a plan based on division interviews. See what tasks are required, and how long they will take for each division.

Put metrics in place. The goal is to determine the efficiency of new systems as a benchmark. Eventually the production of metrics after adoption should show efficiencies of the new system compared to older tasks.

Step 4: Start Small

Training should begin with the BIM manager and a few dedicated associates from the division specified in the implementation plan. The idea is to begin with a small group that can start producing work after their training. The first group's goal is to start using the software and implementing it immediately after training on a project. Unless the use of the software directly follows the training, the associates will forget what they learned.

Five Components of an Integration Plan

An integration plan has five components:

Synopsis This is a brief statement of the company's stance on BIM.

Goals and schedules This section should include all of the following:

- Goals of the BIM integration

- Purpose of the BIM integration

- Team members' responsibility outline; should include new and changed responsibilities

- Software acquisition plan, which should show the following:

  - Training schedules

  - Hardware update schedules

  - Implementation schedule

Additional operational information This includes new contracts and new delivery methods.

Future growth plan This should outline the future goals for BIM at the company.

Supporting articles This should include journal articles, publications, book excerpts, and statistics that make the case for BIM and identify potential opportunities.

When completed, the plan should be compiled into one document and then presented to management. The BIM manager will be responsible for implementing the plan and organizing the training for associates, and it is critical that division managers know that training will be taking project management's time away from their normal day-to-day tasks. Organizing the management of associates and scheduling their training will be challenging, but the rewards, if implemented correctly, are significant.

The next issue involves project choice. Smaller projects provide a scalable way to begin using software effectively, while in a larger project the fee is able to fund research and the purchase of the software. Larger project BIM implementation isn't necessarily to pay for the software; it is to create efficiencies and savings for the project team and the construction company. The size of projects varies, and there are pros and cons to each. This decision will need to be made by the team and will need to be focused on a project where the architect, engineer, and fabricators are all using BIM.

Step 5: Keep the Manager Trained

The BIM manager will need to be trained in all the BIM software that the company uses—not to become completely proficient in all these different pieces of software but rather to gain an understanding of its purpose and be able to competently speak about all the software when requested to report on its implementation. Continuous training will keep the company aware of new technologies, methods, and resources through the manager.

Step 6: Support the Manager by Starting a Department

Implementing BIM in a construction company is in many ways more difficult than in a field such as architecture or engineering. Although an architecture firm might adopt Revit, Bentley, or ArchiCAD, the BIM implementation in a construction company goes through each department and involves multiple pieces of software and overlapping responsibilities. In a typical architecture firm, the role of CAD manager is usually filled by the professional who has been tasked with maintaining firm standards, implementing software, and keeping the licenses up-to-date. In a construction company, the role of BIM manager is specific to each company. Because there is no general consensus about the specific role of the BIM manager and supporting personnel such as BIM specialists, there seems to be a number of companies that have identified that the number of projects within their organizations requires a BIM department. In a construction company, a BIM department should be structured so that the average workload can be distributed effectively among the team. Typically, BIM specialists can run about three to five projects, depending on their experience level, while a BIM manager might be able to handle more. Don't expect to hire one BIM manager and have them effectively run 12 or 13 projects. Think of the construction project manager's project load and staff similarly for the virtual construction department. Because the project manager is responsible for the physical construction, the BIM manager will be responsible for the virtual construction and inform the team about issues before construction on the project starts.

Step 7: Stick to the Plan but Remain Flexible

Possibly the most difficult part of implementing BIM technology at a company is sticking to the plan. This entails supporting the manager, purchasing software on schedule, and making sure associates are being trained in software relevant to their day-to-day tasks. The implementation is successful when the plan is achieved.

Although sticking to the plan is a yardstick for success, it's also important to be flexible. The implementation process can potentially take years, and it's important that the plan stays flexible as new software and other technologies become available and other challenges arise. Software will continually change, so the plan has to adapt to better alternatives that become available as key milestones are reached.

Step 8: Create Resources

Develop internal tutorials and guides. Developing tutorials will help create a reference and a learning point for field personnel, construction management, and other departments. In turn this will create a lean BIM department and the ability to standardize how certain tasks are accomplished. These tutorials may be hosted on a company's website, intranet, FTP or other media for access.

Step 9: Analyze Implementation

Find out how BIM is either improving or not improving processes. Measure to see what components of BIM are realizing the most savings and creating the most value. By measuring where the BIM implementation plan has taken the organization, the manager and the leadership team can gather information and begin to analyze which software is working and where there is room for improvement. It is critical to the success of a BIM division that you avoid pointing fingers. BIM is a growing industry, and certain solutions continue to be tested in the real world. There are so many pieces of software and so many organizations operating with different standards in place that BIM solutions must be customized to complement a company's existing operating platform—that's yet another reason why research is critical, as stated in step 1.

Step 10: Monitor New Software Proposals and Industry Trends

The BIM manager has to constantly be immersed in market trends, new software, and industry publications to stay ahead and aware of industry trends:

- By staying aware of new and emerging solutions, you can begin to develop a plan in your mind that addresses issues at your own company. Constantly question the efficiency of an operation, and continually seek improvement.

- Management will often become interested in what technologies can give them a competitive edge over their competition as more and more owners and clients begin to request BIM technologies. Many companies adopt multiple pieces of software to try to achieve a desired result, but the real market advantage comes with being able to show how a solution has worked (or not) and to learn from the experience.

- This BIM department has the potential to generate revenues outside an organization's bread-and-butter revenue source. One of the advantages of integrating new technology is that by doing so you can create a product that becomes more intelligent and useable by professionals along the path of construction. Sometimes markets are created, just as virtual construction companies have begun to explore what the value is to create a BIM, something that wasn't even considered until recently.

Additionally the BIM Manager should attend conferences, presentations, forums, and construction meetings related to BIM technology to do the following:

- Learn how others are using each piece of software and, in turn, get the message out about the company's experience with these solutions.

- Gather information from these groups and functions to take back to the team.

- Remain aware of new available technologies and get an idea for emerging market trends to make more informed decisions later.

Technology today is moving at an exponential pace. Software development, entrepreneurship, and global communications technologies have created an environment in which being cutting edge requires someone to constantly be informed. A number of online resources are available, such as blogs, content libraries, online model testing sites, and forums. A few are listed here:

www.revitcity.com

www.augi.com

www.bimforum.org

http://bimcompletethought.blogspot.com

www.aecbytes.com

I encourage professionals to share and distribute among their peers the procedures and best practices that educate users about BIM. The dialogue becomes stronger among the BIM community and becomes an invaluable resource to build upon for the present and future generations.

By staying aware of the market and emerging developments, the BIM manager will be able to make more informed decisions about future implementations as well as be able to judge a company's current status compared to the market.

Ten Steps: The Short Version

This sidebar comes from the advice I gave a colleague who was tasked with implementing BIM at his organization. Although it's a humorous slant on the ten steps, it's meant to outline a recipe for successfully integrating BIM into the fabric of a company.

1. BIM doesn't work—people make it work. There is no way you can load BIM onto a machine, plop anyone in front of the machine, and hope that it will somehow make your life easier. In fact, it will make it harder for a while; let everyone know this.

2. BIM is an investment. The easiest way I can explain this is that it's almost like your 401(k) in the form of coordination return. Will you realize the profits immediately? I don't know—probably not. Will you realize your investment six to eight months down the road when you find 188 clashes that equate to more than $2.3 million in change orders? That's closer. Will you realize that investment when you can provide a greater service to your AEC team in improved communication and collaboration? Bingo!

3. BIM will not tie your shoes. I use this phrase in my office when someone thinks that BIM can solve every construction-related problem there is. It's just not true. BIM is still developing. There isn't a "one-software-works-for-everyone-and-will-fix-everything" solution.

4. Start small. A colleague of mine was recently tasked with integrating BIM into his large construction company. He gave me a call and asked me what the best methodology was. He was thinking of training all 16 different satellite offices via web meetings. I told him don't. Start with one office, make it work, and go from there.

5. Train yourself. Make sure you know and learn and continue to learn as much as you can.

6. Start a small, intense training of a BIM team. These will be your disciples and your backbone when you get uber busy. Believe me, it happens.

7. Third, multiply yourself. Create an FTP file where you can put all of the information in your head in the form of tutorials, articles, standards, etc. for everyone to refer to. This will make your life easier as well.

8. Stick to the plan but don't. After you've dedicated three weeks to do nothing but write a plan that includes a schedule and key timelines, and made it generous, be prepared to edit it frequently. People will question why the company is implementing this new strategy. Be prepared to be called *overhead* until you make their day-to-day routine more efficient—and then be prepared to be called *buddy*.

9. Stop and look at what you've done. Get management to review the implementation, and get feedback so you know where to improve. Finally, get some metrics. This will be a little like herding cats, but finding out how BIM has helped or hurt each division will help your decisions.

10. Attend conventions, seminars, and technology expos to learn about what's out there and if it could be helpful to your company. Have a committee that reviews the new stuff and presents a software plan annually to the ownership.

When new technology and software are approved to be implemented, repeat....

Conclusion

BIM is not perfect. It is a relatively new technology when compared with other industry standards. Yet BIM is the greatest technological advance in the AEC industry in this generation. BIM software was developed as a response from design professionals who began to see the need to create a single source of information that can be shared, added to, altered, and responsibly distributed among the design team. We are just beginning to see the full potential of BIM both as a process and as software and what it means to harness its full capabilities. Although these needs will begin to be addressed as the industry acceptance of this new technology and resource grows, everyone must contribute their ideas, criticisms, and suggestions to the industry. We need to understand that as an industry we are a progressive and creative group of professionals who ultimately are playing in the same arena. As construction professionals, we have an obligation to future generations, to the environment, and to improve BIM technology.

BIM and Preconstruction

This chapter explores the beginnings of a BIM process during the preconstruction phase. It delves into the critical path needs of setting up a BIM project, and the aim of this chapter is to show that by laying a solid foundation of responsibilities and goals you can set a project up for success. This chapter includes the following topics:

In this chapter

Planning a BIM project

Contracts

Defining responsibilities and ownership

Digital information transfer standards

Estimating

Site coordination

Planning a BIM Project

A good construction manager knows that any successful project begins with good planning (see Figure 2.1). This is true for planning a BIM project. Although the deliverables are different between virtual construction and actual construction, the goals and focus of completing a successful project are the same. This is why it is important to define how you'll use BIM as a tool before beginning a project. BIM is a great technology and a resource that will continue to grow and change the construction industry for the better; nonetheless, you should approach it with a fair amount of thought in its use. After all, if a mason is just learning how to use a trowel, you can't expect him to build a great cathedral or temple the next day (see Figure 2.2).

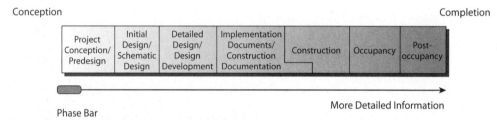

Conception   Completion

Phase Bar   More Detailed Information

Figure 2.1 Planning a BIM project begins before contract negotiation.

PHOTO COURTESY OF BRAD HARDIN.

Figure 2.2 The construction of the Pantheon in Rome required precision, the right tools, and years of experience.

Construction professionals around the world have begun using BIM technology, but there are risks in adapting these technologies on a project, depending on the experience levels of those on the team. Some contractors and design teams expect too much of software or its users and try to use too many tools at once. The group then becomes ineffective and gets bogged down in attempting to understand the array of tools. For example, if the internal goal is to use BIM estimating software on a project for the first time, then it will be difficult to adopt clash detection, animation, and field BIM software for the first time as well. To avoid adopting too much too soon, focus on one area of implementation per project; then dig in deep to find out what worked and to note the lessons you've learned. Treat new projects as opportunities for both new technologies and new experiences, and plan on adapting to these new needs responsibly.

Because each new construction project has so many variables (different project teams, different project types, different locations, and so on), it is difficult to define a one size fits all plan for the use of BIM or to measure its successful use. The best way to produce a successful project is to draw on all team members' experiences while at the same time taking into account the level of sophistication of the BIM user, unique project challenges, and owner requirements. All of these elements contribute to a successful BIM project plan.

The following are critical factors in planning a BIM project:

Educate the team Educate team members on the reasons BIM technology is being used, the desired results, and how the different pieces of software relate.

Achieve team buy-in All the members of the team should realize the importance of their role in the process and have some degree of input throughout the project.

Keep it real Too many times we forget that we are humans dealing with humans, not computers dealing with computers. In this regard, educate teammates on the real-world application and desired physical result of what they are doing.

Set goals This is a good way of quantifying successes in a scalable manner. These goals should be both internal and project focused. A sample goal might be: "Use BIM esti-mating software to run the initial estimate and updates for a project, and quantify the time required for all tasks associated with BIM estimating up to the construction phase." Project focused goals are established in the information exchange and model coordination plans, which are discussed later in this chapter.

Take it a step at a time Identify which technologies your group will be using on a project at what time in a project. These should satisfy both internal efforts and the project owner's requirements.

Select the properly trained staff If it is an internal project goal to familiarize junior project management staff with BIM technologies, then pair them with an experienced user. If the project is highly sophisticated and requires your best and brightest, or outside help, address these issues early.

Challenge the team Adapting and digging deeper into new technologies helps people keep an open mind toward alternative delivery methods, software, and technology and lets all team members have constructive input on a project.

Contracts

The intention of a contract, especially when geared toward BIM users, is not to point fingers if something goes wrong but rather to clearly define tasks, responsibilities, and rights at the onset of a project (Figure 2.3). Unless contract language is defined before creating, using, or transferring BIM technologies into a project, no team member can be held responsible for delivering on their intended goals. And unless the terminology and plan have been established, it is difficult to uphold any standard means of working together in BIM. In a typical BIM contract, there are three groups of professionals: the owner, the design team (architects and engineers), and the contractor. These groups are professionals who share similar but different rights and privileges when completing a BIM-led construction project.

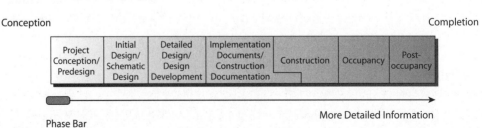

Figure 2.3 BIM contract negotiation begins at the onset of the project.

Contracts on a BIM project determine a contractor's ability to influence and collect and share data throughout the project's life cycle. Thus, it is critical to define these responsibilities early on. BIM contracts in general are in their infancy in the construction industry, and BIM project planning is new ground as well. Current BIM contracts require that the parties entering the agreement have a thorough understanding of BIM processes, model sharing, and ownership privileges. Although this is fine for an experienced BIM team, it can be difficult for a team new to BIM to spell out at the beginning of a project what challenges they are anticipating. If you are a new user, either consult with a professional peer who has entered into a similar agreement or bring on a BIM project consultant to help the parties define what is the best way to distribute roles and responsibilities among the project teams before entering a BIM contract. This will streamline the process significantly and can provide invaluable insight to avoid potential pitfalls ahead. If neither of these options is available, consult with your legal counsel about the contract language, with the focus on integrating BIM and the project team and clearly defining these roles. As this book will show, BIM is most effective when used as part of an integrated effort between the project team. This is particularly evident during project planning when it is being determined when and how BIM is to be used, shared, and analyzed.

A number of groups have model contracts geared toward integrated projects and language you can use to develop contracts. Currently, the American Institute of Architects (AIA), the Associated General Contractors of America (AGC), and the Design-Build Institute of America (DBIA) all have documents that deal with project team integration and alternative delivery methods. However, the AIA and the AGC have to date been the two organizations which have developed BIM contract language that deals specifically with building information modeling roles and responsibilities within the team throughout the construction process. Both of these contracts are similar in appearance to the IE and Model Coordination Plan shown later in this chapter.

Most important in an integrated BIM process is team selection. Ideally, you should select companies and professionals you can work with that have the experience and ability to perform and deliver a BIM project. Effective team selection sets the tone for the entire project, and past successes set a precedent for working well together in the future. Many subcontractors and consultants have no experience using BIM software. In this case, you must find out the stance of the subcontractor. If the subcontractor is receptive to using new technologies and delivery methods, this can be helpful in future projects and relationships. Resource sharing is in the best interest of the general contractor who wants to use BIM in regards to the subcontractor community. Often contractors complete multiple projects with the same subcontractors. For this reason, general contractors who choose to engage a subcontractor to use BIM should develop an information exchange plan between themselves and the rest of the team. In turn, the process of using BIM technology will become more efficient, and experience increases between the two teams and the community at large.

AIA Documents

The AIA has released contract language that addresses the use of BIM as well as integrated project delivery (IPD). These AIA documents require a knowledgeable team of experienced BIM users to define the protocols for sharing, owning, and transferring data throughout a project.

These documents are as follows and can be found on the AIA's website (http://www.aia.org/docs):

AIA A295-2008 General Conditions of the Agreement for IPD

B195-2008 Standard Form of Agreement between Owner and Architect for an Integrated Project

A195-2008 Standard Form of Agreement between Owner and Contractor for an Integrated Project

GMP Amendment to A195-2008 Amendment to A195, defines the GMP and contemplates distribution

Additionally the AIA has contract documentation for the creation of a "single-purpose entity" and uses BIM within the language as well. Per the AIA, "This agreement allows a complete sharing of risk and reward. With this arrangement, owners, architects, and construction managers work together from the beginning to carry out projects with mutually agreed-upon goals and target costs." Although BIM as a technology has gained acceptance, the methods of delivery are new in its approach in the United States, although similar models have gained acceptance and popularity in countries such as Australia and the United Kingdom (www.tradelineinc.com/reports/0A03D1C0-2B3B-B525-85702BCEDF900F61).

Conceptually, the agreement forms an umbrella entity, in which the teams involved are all members. The agreement aims to limit exposure and liability through language that doesn't permit litigation within the newly formed entity. The members are therefore shareholders with the owner, essentially buying a product at an agreed upon cost, which is a building or structure. The professionals providing the services (the architect, engineer, or general contractor) act as the manufacturer of the product. Subcontractors then become the distributors of the product. Upon successful sale of the product to the owner, any profits or efficiencies realized are monetarily rewarded, such as the product was delivered early, came in under budget, or other delivery goals were met.

The document for this type of Single-purpose entity method of delivery is:

C195-2008 Standard Form Limited Liability Availability Agreement for an Integrated Project. (See Figure 2.4)

Figure 2.4 A Project Alliancing Example – National History Museum, Canberra, Australia

The design-build method, as discussed in Chapter 1, has gained popularity in the United States. It is a widely known method of project delivery and a feasible means of delivering a collaborative project. Although the AIA has documents that address the design-build process, there remains the opportunity for further language to be developed within these documents for the integration of building information modeling. The contract language from the AIA A295-2008 document provides an overview of processes and a skeletal outline for how BIM is to work during a project. However, consult with legal counsel prior to engaging in new agreements or attempting to alter typical contracts, because you might find industry-standard language elsewhere. The following are typical contracts for use in a design-build scenario under the AIA agreements:

A141 Standard Form of Agreement Between Owner and Design-Builder (architect as design-build prime)

A142 Standard Form of Agreement Between Design-Builder and Contractor (between design-build prime and general contractor)

A143 Standard Form of Agreement Between Design-Builder and Architect (contractor as design-build prime)

AGC Documents

The AGC was the first organization to market with contract documentation focusing on the use of BIM in a project, on September 28, 2007. These contracts addressed a number of different project delivery methods and the use of BIM. The following are the contracts that were developed, and are available at (www.consensusdocs.org):

ConsensusDOCS 300 For Delivery Methodology

ConsensusDOCS 301 For Electronic Communication and Building Information Modeling

The focus of these documents was to begin identifying key parties and technologies involved in a project and define liabilities, responsibilities, and opportunities "to reflect the project's best interests, rather than a single party's interest" (www .consensusdocs.org/news/20070921-agc.html). The ConsensusDOCS contracts reflect an overall industry interest in a more integrated process and focus on BIM as a tool that can enable the team. These contracts aim to turn the focus away from finger pointing and toward project-focused teams through BIM technology and responsible data sharing and collaboration.

The AGC still maintains contract documents for the design-build method of project delivery. The integration of BIM into these contracts is through the aforementioned ConsensusDOCS 301 BIM addendum. Per the AGC, the BIM Addendum document "is intended to be used as an identical contract addendum for all project participants inputting information into a BIM Model throughout the construction process. The document includes a BIM Execution plan, and allows the project participants to determine the level for which a BIM Model or models may be legally relied upon" (www.agc.org/cs/contracts). In addition, the AGC ConsensusDOCS 200.2, Electronic Communications Protocol Contract, addresses paperless project delivery, how to use information appropriately in transfer and use, and how to structure the IT team. This document aims to address and organize the flow and management of digital data on a project.

The following are common AGC contracts:

AGC 400 Preliminary Design-Build Agreement Between Owner and Design-Builder (review and evaluation of owner's program and development of a price and time to complete a project)

AGC 410 Design-Build Agreement and General Conditions Between Owner and Design-Builder (cost plus with a GMP – Guaranteed Maximum Price)

AGC 415 Design-Build Agreement and General Conditions Between Owner and Design-Builder (lump sum based on owner's program and schematics)

DBIA Documents

The DBIA offers contracts that deal with differing methods of design-build project delivery. The DBIA has not introduced BIM technology or digital information–sharing language to date; however, the most common contracts for a design-build delivery are these (and are found at www.dbia.org/pubs/contracts):

DBIA 520 Preliminary Agreement Between Owner and Design-Builder (review and evaluation of owner's program and development of a price and time to complete a project)

DBIA 525 Standard Form of Agreement Between Owner and Design-Builder (lump sum)

DBIA 530 Standard Form of Agreement Between Owner and Design-Builder (cost plus fee with an option for GMP)

Defining Responsibilities and Ownership

In all the types of contracts discussed, the project team members need to define how data will be shared and used (see Figure 2.5) in order to do the following:

- Eliminate confusion

- Organize tasks

- Standardize information transfer

- Define the schedule

- Focus on project quality

Anticipation is always a better approach than reaction; just ask a goalie. To carry out a BIM project, you need to create two plans at a minimum that anticipate some of these issues. The first is the Information Exchange (IE) Plan. The second is the Model Coordination Plan. Both plans should begin as drafts and then be reviewed and approved by the team in preliminary contract meetings. These two plans or other similar plans will then be carried forward as an addendum to the BIM contract language for the project. Just as the owner, architect, and contracting team must define critical deadlines, goals, and methodologies, you need a road map for the BIM portion of a project as well (see Figure 2.6).

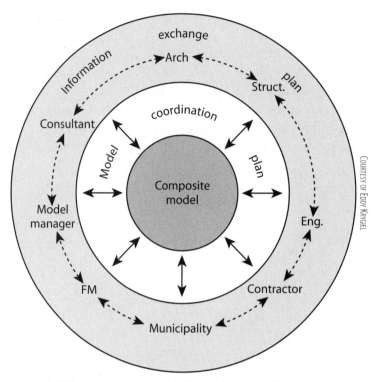

Figure 2.5 BIM has multiple uses and stakeholders, so defining the rights and responsibilities are critical between team members and model users.

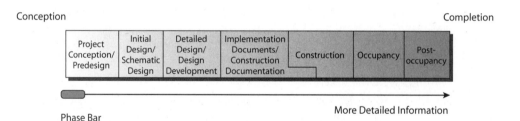

Figure 2.6 Information and model exchange plans are established at the beginning of the project.

There are a number of legal questions arising from the use of BIM, which raise concerns as to a professional's exposure when utilizing BIM. A question of particular concern is: who is held liable when inaccurate information is input into the model?

This is particularly relevant to a process that involves any type of bidding. Depending on the type of contract, this topic is not too much different from existing inaccurate or incomplete CAD derived data and is usually covered in the E and O (errors and omissions) portion of the contract documents. Ultimately, it is the responsibility of the licensed professional to verify that accurate information is input into the model, just as CAD or line based documents. BIM is unique in that those terms have yet to be clearly defined as to what information and level of detail (LOD) is expected from each team member. The model coordination plan aims to clarify expected BIM deliverables along a project's milestones. Additionally, it should also be outlined in the IE plan that it is the responsibility of the team to notify other players to inaccurate model information, which helps in coordination. The creation of the "perfect model" in a project is often a moving target as the design changes and shifts to accommodate this additional coordination. It is interesting that the ability to add and coordinate documentation above and beyond what is currently accomplished brings to light *additional* liability issues, when in reality the documentation is typically more accurate than before. In part, this is due to the relatively little amount of litigious information about BIM, which ironically somewhat hinders the development of the language associated with better technology, process, and information sharing.

Another question often asked is: who owns the model? This question requires entirely different thinking because the answer should be that no one does. If the library has all of the information you need to efficiently share and borrow information, then the actual ownership of the library becomes somewhat of a non-issue (provided the user has some means of access to the library). This is not to say that there is no need for a gatekeeper or model content manger; the reality is that the need for this person is paramount when dealing with a composite modeling strategy. Current contracts give the model manager role to the architect. However, this might not always be the best solution, as architects may be new to the use of BIM and have more experienced teammates who would better manage this coordination. Additionally, some projects may have a certain focus where a structural engineer has the majority of input and wants to shoulder the bulk of the coordination, such as a bridge project. Lastly an owner might provide their own model coordinator or require the contractor to be the model coordinator on projects that require complex phasing or construction methods. Although the answer to the model sharing solution does not have a standard answer, it does offer flexibility for the team and positions the responsibilities of a project to be most effective (see Figure 2.7). The need for this plan to be clear at the beginning is critical to its success.

Schematic Design Model Coordination Draft Plan

Model Portion	Responsible Team Member	Control Type (Complete, Partial, None)	Special Conditions	Comments
Exterior Walls	Architect	C	Pre-cast Panels	Panel sizes need to be reviewed by Structural
Structure	Structural Engineer	C	Steel beams, joist system, and interior bearing walls by engineer	Reference structural model to be . RVT format for linking
Doors	Architect	C	None	None
Storefront Glazing	Architect	C	None	Million spacing to be reviewed by structural at 100% SD's
Floors	Architect	P	Verify depth and floor level with Structural model	
Floor Structure	Electrical Engineer/ Subcontractor	C	Structural .RVT model to be linked at 75% SD's	Verify edge of slab conditions and sufficient edge clearance
Roof	Architect	C	Membrane roof system design by architect	Verify bearing height of roof on structural model
Roof Structure	Structural Engineer	C	None	Sizing TBD
Footings	Structural Engineer	P	Concrete footing system–check alignment with Architect. Foundation walls	Thickness and design to be verified by Architect
Foundation Walls	Architect	P	None	Verify bearing height of model with Civil model
Electrical	Electrical Engineer/ Subcontractor	P	Lights and AV in ceiling and wall by Architect	Electrical reference model to be in .RVT format for linking
Plumbing	Plumbing Engineer/ Subcontractor	P	Toilets, sinks, and showers by Architect	Plumbing reference model to be in .RVT format for linking
Mechanical	Mechanical Engineer/ Subcontractor	P	Supply and return air vents by Architect	Mechanical reference model to be in .RVT format for linking
Civil	Civil Engineer	C	Parking model to be provided by Architect for use in DD's	Civil reference model to be in 3D .DWG format for linking

Figure 2.7 The model coordination plan is relevant to the information being input into the model and the information exchange plan is relevant to the exchange of information between team members.

Information Exchange (IE) Plan

The following is a preliminary draft of the IE plan for different members of the project team:

Architect's IE Responsibilities

- Responsible for communicating the design intent of the structure through documentation (both real and virtual) in agreed upon manner (FTP site, Newforma file transfer, DVD) to the team.
- Responsible for coordinating information regarding life safety, code compliance, and accessibility issues
- Responsible for issuing BIM information and documentation to MEP, structural and civil engineers, and other consultants throughout the design process at agreed upon project milestones.
- Responsible for tracking the date, time, and people to whom the design documentation was transferred; includes FTP site uploads for consultants and engineers

- Responsible for model coordination and model ownership through the design phase and 100 percent construction documentation milestone of the project
- Responsible for owner-client (facility model) BIM through the design phase and 100 percent construction documentation milestone of the project
- Responsible for submitting responses to RFIs in a digital format (PDF, DWG, .RVT) in tandem with or in place of paper submittals
- Responsible for submitting punch list and project closeout document in agreed upon format (PDF or Vela System) for owner's, architect's, and contractors' use
- Responsible for submitting all construction documentation, specification, warranty, BIM, and other design information in agreed upon digital format at project closeout

Contractor's IE Responsibilities
- Responsible for reviewing the architect's design intent model and assigning a budget to the BIM using the agreed upon software (Innovaya, Constructware, Vico Estimator)
- Responsible for maintaining and layering information onto the singular model as defined and agreed upon (URL data, unused data fields for coordination, embedded worksets, sequencing and date information, assembly code estimate linking, specification information, and custom fields if not provided)
- Responsible for creating and maintaining sequencing animation throughout the project, beginning at 100 percent schematic design submittal
- Responsible for clash detection reporting on a biweekly basis beginning at 50 percent design development submittal; creates and maintains a clash resolution report from design development through project closeout
- Responsible for digital RFI issuance and utilization of the agreed upon software and transmittal system (Prolog, Adobe, NavisWorks)
- Responsible for creating the as-built or record BIM from the beginning of construction to the completion of the project and delivery to the owner and architect as agreed upon; includes all as-built and site alterations and changes

Mechanical, Electrical, Plumbing, and Structural Engineers' IE Responsibilities
- Responsible for the 3D creation of the systems in the design team's preferred software per the requirements set forth by the model coordination plan and exported in the agreed upon format (DWF, 3D CAD, IFC or other) to the contractor to be used in the clash detection process
- Responsible for accurately updating the project team on model changes
- Responsible for making recommendations to the design team on performance and manufacturing efficiencies in their respective system designs

- Responsible for making alterations to the design model after a clash has been reported within one week of the report being issued
- Responsible for submitting the completed clash report to the contractor in tandem with completing the model alteration
- Responsible for delivering all 3D documentation to the contractor at project closeout for the creation of a record BIM

Although these are some simple draft examples of what the IE responsibility matrix should contain, it's a good start in working toward your project-specific plan. This responsibility matrix typically will be completed by the architect and construction manager, because these two parties will be responsible for the majority of information sharing and management. The engineers and subcontractors on the team should then review this document with regard to their respective scopes of work. If the delivery method does not permit some of the relevant parties to participate early in the process, be sure that these members are acutely aware of the responsibilities associated with working on the project team as early as possible. Integrating the technology and goals in the RFQ as part of the project deliverables further defines to the rest of the team members what technology will be used and for what purposes. Communicating early and choosing the right partners from the start of a project lays a strong base to build upon and limits confusion in the project.

Model Coordination Plan

Although the IE plan responsibility matrix defines responsibilities for the larger tasks, the team should also develop a Model Coordination Plan, sometimes called the *BIM protocol document* or the *BIM guide*. This document spells out who is responsible for the development and analysis of the model at what point in the project's progress and to an acceptable level of detail. In some projects, the design team might have management privileges of the model throughout the entire process. In others, the design team might hand the model off to the construction manager to update the model as the project is being constructed so the construction manager can verify changes and on-site design alterations, as in the earlier IE plan example. This is also common when the project calls for a completed record BIM to be delivered to the owner at the end of a project; it streamlines the work that the architect must do during construction. In other projects, the owner's representative might maintain the model and track the transfer and compilation of data through a virtual construction manager. This position was unheard of a decade ago, but it is becoming more and more popular as a project's size and complexity increase and timelines continue to become tighter and more sophisticated. In all of these options, there must be a clear means of transferring the data at critical timelines and an understanding of who will be responsible for analyzing that information along the construction process. This is the purpose for the Model Coordination Plan.

A Model Coordination Plan spells out which team member is responsible for which portions of the model throughout a project. In the past, projects have lacked sufficient information sharing because one or two members on a project team had proprietary information in the model prior to transferring. Although it's a great idea for firms to develop these libraries of assembly and component information to improve internal efficiency, doing so should not inhibit the progress of the project and become more of a liability to the team than a resource. The information in a BIM project is different from a CAD project, because BIM information is intended to be routinely shared among team members. If required, a nondisclosure document may be signed at the beginning of a project if a certain organization is worried about transferring legacy information.

Typically, Model Coordination Plans begin at the onset of a project and defines in detail the model responsibilities from start to finish. Table 2.1 is an example of a draft model coordination plan for schematic design.

▶ **Table 2.1** Schematic Design Model Coordination Draft Plan

					Model Portion	Responsible Team Member	Control Type (Complete, Partial, Type)	Special Conditions	Comments
A	Substructure	A10	Foundations	A1010	Foundations layout	Structural	C	Structural .RVT to be linked at 75% SD's	Verify alignment with foundation plan
				A1020	Special foundations	Structural	C		
				A1030	Slab on grade	Architect/civil/ structural	P	Verify bearing height w/ civil	Verify thickness with structural
		A20	Basement	A2010	Basement excavation	Contractor	C	Preliminary Formwork, equipment and laydown plan in .RVT	Contractor to compile Navis file at 100% SD's
				A2020	Basement walls	Architect	C	Coordinate w/ structural at DD	Verify approx. thickness with structural
B	Shell	B10	Superstructure	B1010	Floor (from steel up)	Architect/structural	P	Verify depth and floor level w/ structural	Verify floor assemby w/ structural
				B1011	Floor structure (from concrete down)	Structural	C	Structural .RVT to be linked at 75% SD's	Verify edge of slab conditions and sufficient edge clearance
				B1020	Roof structure (from roof down)	Structural	C	Structural .RVT to be linked at 75% SD's	Approximated depth for structure required to architect
				B1021	Structural walls (internal)	Architect/structural	P	Structural .RVT to be linked at 75% SD's	Coordinate layout w/ architect
		B20	Ext. Enclosure	B2010	Exterior walls	Architect	C	Precast panels	Panel sizes need to be reviewed with structural
				B2020	Exterior windows	Architect	C		
				B2021	Exterior curtain wall	Architect/structural	P	Coordinate height w/ structural	Verify mullion spacing with structural at 100% SD's
				B2022	Exterior storefront	Architect/structural	P	Coordinate height w/ structural	Verify mullion spacing with structural at 100% SD's
		B30	Roofing	B3010	Roof	Architect	C	Membrane roof system design by architect	
				B3011	Parapet, flashing, other	Architect	C		
C	Interiors	C10	Interior	C1010	Partitions	Architect	C	All partitions by architect	Locate fire rated walls for 100% SD's
				C1020	Interior doors	Architect	C	All doors by architect	Locate fire rated doors for 100% SD's

The Model Coordination Plan should break out timelines and milestones, specific portions of the model, who is responsible, control type, special conditions, type of software and project file type, and comments, at a minimum. Often there is a need for further information and category breakdown, especially on complex projects with multiple phases or existing conditions or where level of detail must be defined. The Model Coordination Plan supplements the information exchange plan as a detailed outline for information being built into the model and as a tool for coordination among the team members. It is impossible to address all the conditions that arise through the development of the project. However, the model plan aims to make it easier to determine who is responsible for certain elements that come up, as opposed to not having a plan at all. Ultimately, the plan should be organic—developed, updated, and submitted before the completion of any project milestones. This maintains a clear picture of responsibility and eliminates much of the confusion that can occur with the development of the BIM.

Both of these plans are critical to drawing boundaries and defining relationships in a successful project. However, these plans must allow for flexibility; they have to account for changes and for different team members needing differing levels of access to information throughout a project. For instance, the level of access that a structural engineer might need in schematic design or initial concept development might be different than anticipated. Therefore, the responsibility level might need to shift during the construction document or implementation documentation stage. Buildings and structures come together organically. No continuous level of detail exists for each stage of a project for each team member. For example, a change in the design from a punched window facade to four stories of curtain wall will require the mechanical engineer to redesign the mechanical systems to work with the new concept. This additional work puts any deliverable for the mechanical design behind where it would be if the engineer were proceeding with the original layout. For this reason, the level of detail isn't as critical as that team members provide the information needed in time to further the entire design.

Digital Information Transfer Standards

Transferring data among team members can be a time-consuming process. In the past, architects have transferred data to the contractor by means of printing CAD files or other 2D drawing information into a PDF or other noneditable format and then mailing CDs, posting to a firm FTP site, couriering DVDs, or attaching ZIP files to email. This followed converting the native file format and limiting any input from other team members into the native file. Until recent years, many projects were architect-led, and the architect dispersed the data to the relevant parties. But more recently, some projects are contractor-led, and the contractors are running more sophisticated technology than the architect's software. Although this is not always the case, it is definitely an interesting turn of events in that both professions are realizing the value of information-rich data and are demonstrating this belief by investing in BIM software. With collaboration and integration in mind, many companies are now seeing the value in creating information sharing standards before construction. (See Figure 2.8) This concept is different from past models in that the goal is not at any point to "lock" or "freeze" data prior to distribution. The freezing of data occurs, for example, when a sheet is plotted or is converted to another read-only format. BIM involves linking multiple models, testing them, and then further coordinating the virtual construction. Because there are so many stakeholders and so many model changes and updates, an archiving strategy must be developed. Archiving is useful for many reasons, such as when certain project milestones must be looked at to gain an understanding of completion, or with previous design changes, or to follow the cost estimate history. Additionally, this is needed in a BIM-focused project not only for milestone reviews but also as a means of backing up previous data or design options.

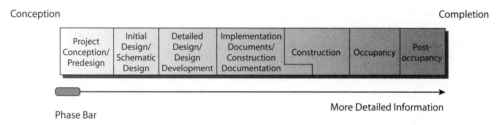

Figure 2.8 Project transfer standards need to be complete at the time of contract negotiation.

BIM projects are only as useful as the last model update. If the project is to be a BIM project, the project needs to stay in BIM models. Architects and engineers new to BIM might be tempted to export the model to CAD at the end of a project. If the model is exported to a CAD drawing format and the BIM technology is abandoned, the file loses the intelligence and the advantages of using BIM software to begin with.

There are two schools of thought on using BIM:

- Creating a parallel BIM model for use by individual professionals
- Using a composite BIM model for all disciplines

Parallel BIM

The idea of parallel BIM model is to create a separate model for use by contractors based on the information provided by the architects and engineers. Although this helps ensure that the model will be useful and specific to the contractor's needs, there is a liability that the model may not be an exact reflection of the architect's and engineers' models; there may also be a conflict with the architect's and engineers' contractual responsibility for communicating the design intent for the project. Allowing the contractor to create the model independently from the design team increases the risk on a project. The architect's design intent can become decoupled from the contractor's model as the contractor continues to edit it. But the architect is responsible for the design and life safety of the project, as are the engineers who sign and stamp their designs. In this redundant modeling strategy, contractors expose themselves to litigation by creating a model that might differ from what was designed by the architects and engineers.

Some construction management companies promote parallel modeling as a best practice in dealing with collaborative BIM projects; they miss an opportunity in working together with a project team, and they fail to look at solutions in the software. There are ways to embed, extract, and analyze information in BIM without creating an additional model. Some special cases exist where a temporary parallel model is a necessity, but overall the industry is moving toward an open, usable tool as software becomes more sophisticated. The practice of creating silos of information that are independent and separate from each other will continue to diminish.

Single BIM

The second method of BIM is maintaining a composite BIM model. Although this method requires an advanced understanding of BIM software programs and relationships, it is truly where BIM shines. This book explores new means of transferring data in a manner that supports the case for a single model.

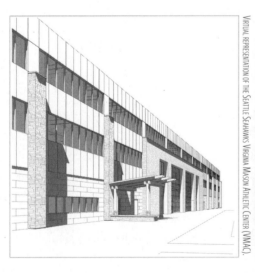

VIRTUAL REPRESENTATION OF THE SEATTLE SEAHAWKS VIRGINIA MASON ATHLETIC CENTER (VMAC).

SEATTLE SEAHAWKS VMAC. PHOTO BY BRAD HARDIN.

Understand that singular modeling is not necessarily everyone working on the same model at the same time; instead, it lets users work on their own models and link or import the models together to create a "composite" model. Users from around the world or in the same room create and build BIMs, which can then be linked into a single model for estimating, clash detection, sequencing, and other analysis. The two software programs that this book uses to accomplish this are Autodesk Revit and Autodesk NavisWorks.

Team Communication

Using either a parallel or composite modeling approach, a project team must establish a means of transferring information. This is the basic purpose of a number of tools. These tools can include an FTP site, software that integrates and tracks file transfers such as Newforma, or an extranet. As more tools are made available in the industry every day, a general understanding of how to transfer files is good to have when creating a standard.

FTP

File Transfer Protocol (FTP) is a means of transferring data directly from a server to a user who has access to it. This is available for any Transmission Control Protocol (TCP) and works across different operating systems.

These are some of the advantages to using FTP with project team members:

Ease of use Users copy from their hard drive and paste onto shared server space.

Centrally located information Information is contained in one place and available for access and review at any time.

Minimal software required An Internet connection and a browser are usually all that is required.

These are some of the disadvantages:

No tracking capabilities in place Users can't verify whether a file was successfully downloaded by other users or by whom it was downloaded.

Security maintenance Keeping track of which users have access to which data can be cumbersome, and password sharing limits security.

Manual archiving Archiving consists of taking older files and creating archives of them, typically in dated folders, for any future use.

Many firms use FTP; overall, it has proven itself an adequate, though time-intensive, means of transferring data.

Newforma

A software product that shortens the time it takes to post and transfer model data is Newforma (www.newforma.com). Newforma is a client-server system with tracking and information management capabilities. This robust tool uses Microsoft Outlook to send files and tracks information that is missing in an FTP-only system.

These are some advantages to using a Newforma system:

Controllability of file transfers Newforma lets users decide who should get relevant information as opposed to sending all data to the whole team.

Tracking file transfers Newforma tracks the date and time and whether the file was successfully downloaded for the user, creating a log of this information for the user.

Searchable content Users can search for data sets distributed in the past, and they can generate additional required copies, using the archive feature, which creates and stores all files sent for a project.

These are some of the disadvantages:

Expense of software Although the system is robust, users pay for it.

Learning curve Using new software and installation can take time.

Newforma is a robust tool that focuses specifically on the AEC industry. The company has begun to look at integrating its software for use with BIM projects (www.aia.org/aiarchitect/thisweek07/1019/1019rc_face.cfm).

Extranet

Lastly, an extranet is a private network that focuses on business-to-business data transfer without access from the public on the Internet. As bandwidth increases and more and more data is stored on web servers, an extranet might be a tool to consider.

Using an extranet provides these advantages:

Security and privacy Through the use of firewalls, digital certificates, and encryption, it offers a high level of security.

Exchange of large volumes of data Using electronic data interchange (EDI), large amounts of data can be accessed and downloaded quickly.

Usefulness for larger groups Extranets allow large numbers of users to access private information relevant to a project team.

Here are some disadvantages:

Expense Hosting and maintaining an extranet internally can be costly.

Access to all data Instead of specific access, generally extranets provide users with access to all data within the extranet.

> ### Working Directly Across The Internet – Or In The Same Room
>
> As technology and software develop, users may be able to directly link to a single model and work through the Internet. In the meantime, the options described here are useful for transferring information among team members virtually. Another solution is to have all team members work from the same room on the same network to compensate for data transfer rates. This introduces a level of efficiency that would not be possible in a virtual sharing environment.

Developing the project standard begins during contract negotiation. The team then agrees upon a method of transferring data for the project. This involves posting the data, letting relevant users know about the post, and archiving past data. The information should be easily accessible for all parties from subcontractors to owners, and the process of transferring this data should remain simple. Whichever system is agreed upon by the project team, it is a best practice to include this in the model coordination plan.

Estimating

Extracting quantities, areas, and volumes from a model is one of the most useful functions BIM technology offers. The following tutorial uses a Revit model and shows how to use Innovaya's estimating tool to extract quantities from the model and then tie them to a Timberline estimate. Other software combinations are also available, such as Vico Software and Beck Technology's DProfiler, with which you can accomplish similar tasks.

Estimating takes place during preconstruction (see Figure 2.9). The procedure that follows shows how to complete updates. This example assumes that all you have received is a business development model that is in its infancy. You start by gathering square footage data and then begin to link model objects with Innovaya for future updating.

The examples in this book progress in level of detail and become more advanced, simulating an actual construction project. This lets you use the tools as you might encounter them in the course of a typical construction project.

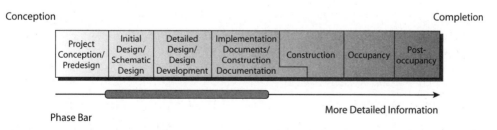

Figure 2.9 Project estimating continues from schematic design to construction.

Revit

Autodesk Revit (www.revit.com) is a BIM software modeling program that allows you to design with parametric modeling and drafting elements. In other words, the model is interconnected; a change in one place propagates changes throughout the model. For instance, if a wall is moved 3 feet in plan, that wall now moves in elevation, section, perspective, and all other relative views. The concept of a single database file that can be shared among multiple users is unique to BIM and separates it from isolated CAD drawings. When a CAD drawing changes, all relevant views must be altered to maintain document accuracy. Revit is not the only BIM software available. ArchiCAD, Bentley, and VectorWorks, among others, all offer BIM modeling packages, which all accomplish approximately the same tasks. A good reference to compare these different pieces of software is the *BIM Handbook: A Guide to Building Information Modeling for Owners, Managers, Designers, Engineers, and Contractors* by Chuck Eastman, Kathleen Liston, Rafael Sacks, and Paul Teicholz (Wiley, 2008), which does a comprehensive job of showing what programs are available and for what purposes. By contrast, the book you hold in your hands features tutorials and walk-throughs using some of these programs. (It's beyond the scope of this book to offer tutorials on all the BIM software packages available.)

The first file you'll use is example-core-shell.rvt. This file, and all of the other tutorial files used in this book, can be found on the website: www.sybex.com/go/bimandconstruction. This is a business development model that reflects a schematic concept, where the architect might have established the basic building program and for which the structural engineer has begun sizing structural members for floor-to-ceiling height reference and initial coordination. When you examine the model in Revit, notice that no mechanical, plumbing, or electrical information is associated with this model; it is merely a schematic design with which the engineers might begin their calculations. You will start by tackling an initial estimate for the core and shell model.

Opening the Model

1. Open Autodesk Revit 2009 (see Figure 2.10), and choose File > Open.
2. Navigate to your CD drive, and open example-core-shell.rvt (or download it from the book's companion web page, www.sybex.com/go/bimandconstruction).
3. Once Revit has loaded the file, click the View tab, and specify 3D view or click the 3D button at the top of the Revit toolbar (see Figure 2.11).

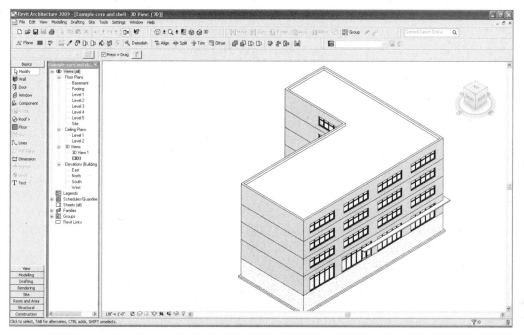

Figure 2.10 Revit user interface

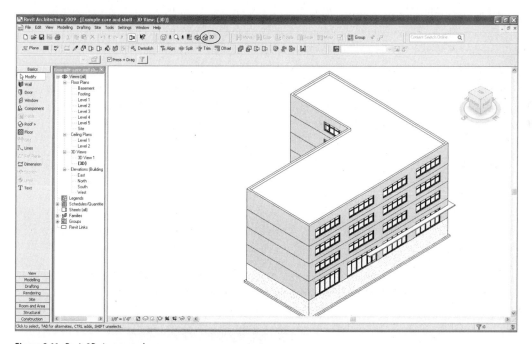

Figure 2.11 Revit 3D view control

4. With the 3D view open, you can move around and position the model for viewing and for enhanced understanding of the model, giving you a better understanding of the scope of the model and showing you any potentially incomplete components around the structure.

   To orbit around the model, click any component in the model and then hold the Shift key and the middle mouse wheel button down to orbit around the selection. This is the easiest way to orbit. To pan in Revit, hold just the middle mouse wheel down and move the mouse. Scrolling the wheel zooms in and out of the 3D view.

5. Click View > Visibility/Graphics to isolate components. Alternatively, type **VG**.

Once you have explored the sample structure enough to have a basic understanding about what is being estimated, you can export the model into an Innovaya (INV) file.

Which Way Is It Going?

Bidirectional model linking is different from model exporting. Model exporting is when BIM information is taken out of its native file format and made available for another program to use. Although the model is providing the information, once the model is exported, there is no way to input information into the new format and then update the model. With bidirectional linking, the information can flow and expand between the software tools.

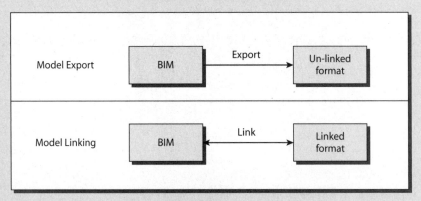

Adding and compiling information in multiple programs as the model is being developed is the most effective means of taking off model components and updating them throughout the course of the project. This is also a goal of the National Institute of Building Sciences (NIBS) National Building Information Modeling Standard (NBIMS) and the International Alliance for Interoperability. BIM shouldn't be limited by proprietary software; compliance with industry standards can maximize flexibility and productivity in the future.

Innovaya Composer

Innovaya Composer (www.innovaya.com) is a model-linking Revit plug-in tool that converts an RVT file to an INV file. In essence, it converts all the BIM components into a file format that is easy to view as estimate-related data and quantity information. Innovaya Composer creates a link in the chain from Revit through Innovaya to Timberline.

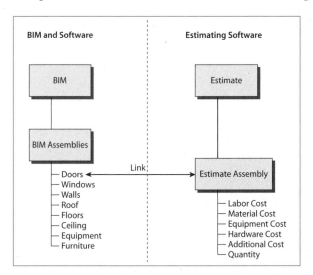

In Revit, choose Tools > External Tools > Innovaya Composer for Revit (see Figure 2.12).

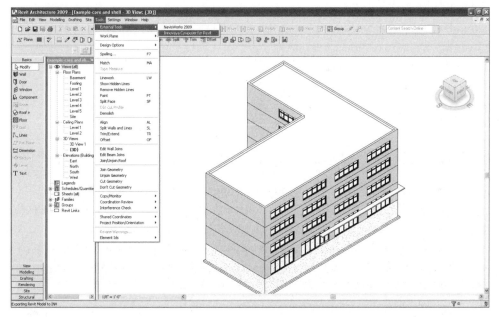

Figure 2.12 Exporting to Innovaya

This opens the Innovaya Composer for Revit dialog box (see Figure 2.13).

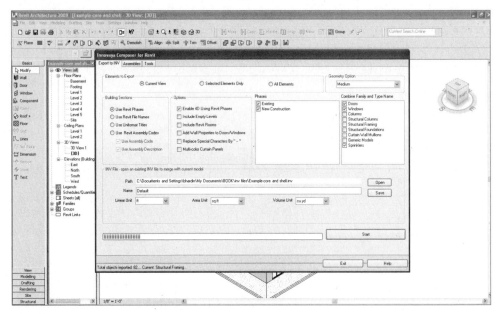

Figure 2.13 Innovaya Composer for Revit dialog box

The dialog box has three tabs: Export to INV, Assemblies, and Tools. The Export to INV tab includes the Building Sections group with the following four options:

- Use Revit Phases
- Use Revit File Names
- Use Uniformat Titles
- Use Revit Assembly Codes

These options specify how you want the components in the model to be categorized. The order of appearance will depend on how you specify the data to be compiled. For this exercise, select Use Revit Phases. At other times, you may prefer Use Revit File Names or Use Uniformat Titles. The Use Revit Assembly Codes option is rarely used, because it categorizes the components into assemblies as defined by the Revit software and isn't typically the best tool to use for interoperability.

The Options group allows you to select one or more of the following preferences:

Enable Revit 4D Phases This enables sequencing videos completed in Innovaya's Visual 4D Simulation to be exported. These videos animate the model components tied to a phase in the Innovaya software. For the sequencing video in the example, I will use NavisWorks.

Include Empty Levels This function lets the estimator assign a cost to a particular level as a line item in Innovaya.

Include Revit Rooms This function lets the estimator break out the rooms into room types to be assigned a cost as a line item.

Add Wall Properties to Doors/Windows Selecting this box adds the area, volume, and other characteristics typically associated with a wall to door and window information.

Replace Special Characters by "—" This allows you to replace any special characters.

Multi-color Curtain Panels This enhances the view of curtain panels.

The Phases area lists the current project phases. You can check all or only new or different phases, as needed, for export.

The Combine Family and Type Name area lets the estimator select the family and type names to combine so there is only a type name in the Innovaya file.

The dialog box has these additional controls:

Path This shows the full path where the INV file is to be saved.

Name This allows you to create a name for the INV file.

Linear Unit This allows you to choose the linear unit setting for the file.

Area Unit This allows you to choose the area unit setting for the file.

Volume Unit This allows you to choose the volume unit settings for the file.

One of the great features of Innovaya is that you can merge multiple models into a single file. This is effective when one portion of a design has been completed and one is still in development.

Next you'll export the model using the default settings.

Exporting an Innovaya File

1. Click Start at the bottom right of the Composer window. This opens the Specify an INV File dialog box.

2. Specify where you want to save the file. Typically, it is best to assign a date and keep all the INV files in one folder for future use.

3. Click Save. You should see a message that your export was successful (see Figure 2.14).

You can use Innovaya Composer to merge multiple Revit model files into one INV file.

Merging Multiple Revit Models into an Innovaya File

1. Open an existing INV file that contains previously exported Revit models by clicking the Open button.

2. Click Start at the bottom right of the Composer window. This opens the Specify an INV File dialog box.

3. Specify a new location where you want to save the file.

4. Click Save. You should see a message that your export was successful.

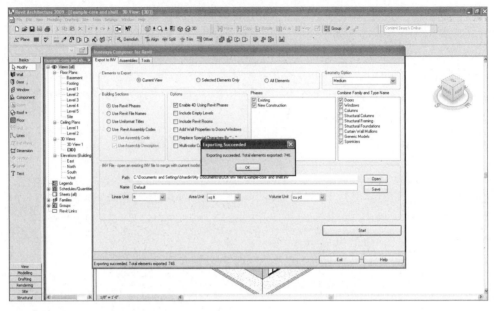

Figure 2.14 Successful export message

With Innovaya Composer, you can also synchronize design changes between two editions of a Revit model.

Synchronizing Design Changes in a Revit Model with an Innovaya File

1. Open an existing INV file that contains previously exported Revit models by clicking the Open button.

2. Click Start at the bottom right of the Composer window. This opens the Specify an INV File dialog box.

3. Keep the save location unchanged.

4. Click Save. You should see a message that your export was successful.

Innovaya

Innovaya Visual Estimating (Timberline) is not a modeling platform; rather it is a way of displaying model types and elements and assigning costs to them. Figure 2.15 is the introductory Innovaya screen.

Innovaya Visual Estimating acts as a connector between BIM software (Autodesk Revit) and estimating software (Timberline and MC2). Using BIM elements with assembly codes, Innovaya sorts the elements and compiles their information into *managed quantities* (MQs). These are different ways of grouping components into levels, types, phases, and so forth. One of Innovaya's biggest strengths is its ability to maintain *sticky* memory for model objects that have been linked, or *pathed*, to Timberline assembly costs. This vastly improves the takeoff process when compared to taking off the building again and again as the model and design changes using On-Screen Takeoff or a digitizer. Typically, the accuracy is better as well, but as in most things in a BIM project, the takeoff is only as accurate as the model. Incorrectly modeled elements will be taken off as modeled. For example, walls that are modeled to the bottom of deck, but are only 8′ 0″ high with kickers will be reflected inaccurately in the estimate. Keep in mind that the software is only relaying the information in the BIM. This is why it is critical to input information as accurately as possible when creating the model.

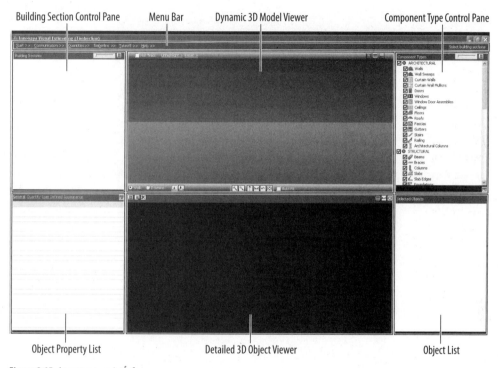

Figure 2.15 Innovaya user interface

Exploring the Innovaya Interface

1. Launch Innovaya Visual Estimating (Timberline).

2. Click Start > Open Project, as shown in Figure 2.16. Innovaya should default to the location you last exported a file to. If not, specify the file you exported.

    The Component Types pane at the top right is now populated; it shows that walls, doors, and windows are loaded into the file. The Building Sections pane has Existing and New Construction tabs; if there are multiple phases to be estimated, it is usually best, for clarity, to export only the relevant phase or to isolate the phase in the INV file.

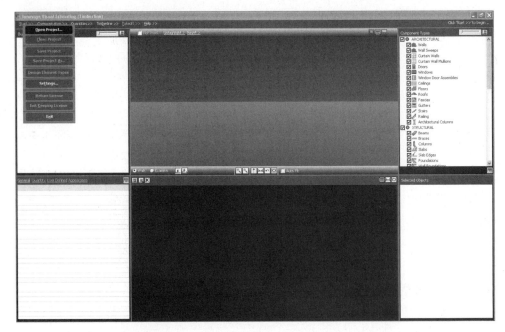

Figure 2.16 Opening a project in Innovaya

3.   Maximize the New Construction window at the top left. The sections are
        divided into levels. Note that the zzzUnassigned category helps you verify the
        completeness of the takeoff, because it lists components that do not yet have a
        cost assigned.

Estimating the Blob

One of the great values of BIM estimating is the ability to assign a cost to every item in the
model. By isolating the items that don't have costs assigned to them, you can make sure your
estimate contains all the items in the current BIM. So, what do you do when you can't figure out
what the item you're estimating is?

This happened to me on one of the first projects I estimated. I kept isolating the unassigned
items (zzzUnassigned category in Innovaya), and I kept getting what looked like a giant blob in
my model. At first I thought it was just a software error, and then I found out that the blob was
an actual object that someone had taken the time to model.

I brought up the blob in our next progress meeting only to find out that one of the younger
interns on the architectural team had modeled a representation of the sculpture that was to
occupy the lobby space. At first we laughed about it. Then we found out that the base of the
sculpture was in the base building scope and that we needed to budget for the material and construction of the base. So, it ended up being a good thing he had modeled it!

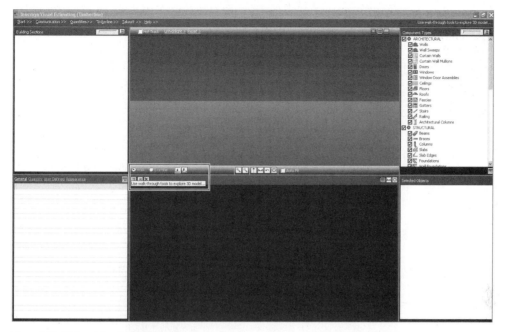

Figure 2.17 Walk and Examine tools in Innovaya

4. In the middle of the Innovaya window (see Figure 2.17), select the Walk option.

5. With the Walk option selected, note the two icons just to the right of the Examine option:

Walk The first icon maintains a uniform perspective height. Click and hold the left mouse button and move the mouse to virtually walk through the model.

Pan The second icon lets you pan the view around and look up and down to better understand ceiling and floor elements.

6. With the Examine option selected, you can change the perspective height and angle. Note the three icons to the right: Spin, Roll, and Pan. These are self-explanatory. Click and hold the middle mouse wheel button to manipulate the model in any of these modes. This may take some getting used to, because these controls differ from those in Revit.

Explore the other controls in the Innovaya window at your convenience. Keep in mind that some buttons are toggles that you click once to activate a mode and again to exit that mode.

Innovaya lets you specify what items you want to see in your estimate through managed quantities. You can specify MQs on the Quantities tab. There are multiple options for generating, saving, and managing these quantities. However, just because you can see the items in the view doesn't necessarily mean the items are in the MQ.

The user uniquely specifies MQs; they allow the estimator to take off different arrangements of quantities without having to save multiple files.

Generating Managed Quantities

1. Select Quantities > Batch Generate (see Figure 2.18). Batch Generate is a tool that generates an MQ file for quantities in the model. The default is all the items in the BIM; however, it is possible to specify categories such as walls, ceilings, and doors and then generate quantities for those items.

   This opens a dialog box (see Figure 2.19) that allows you either to create an MQ file for all the items in the BIM (default) or to specify a group of items that you would like to take off.

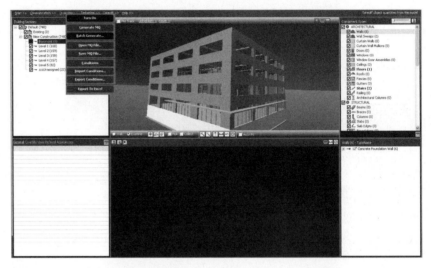

Figure 2.18 Using the Batch Generate function

Figure 2.19 Batch Generate Managed Quantities (MQ) dialog box

2. Use the default settings, and export all the items in the model. To do this, verify that all the items in the Component Types pane are selected, and then click the Generate button.

   At the bottom left of the main Innovaya window is now the Managed Quantities pane, listing the items you selected. You can sort the items by building section or by component type by clicking the corresponding button. For example, if you are estimating a large high-rise condominium project, you may want to see a breakdown of the items in the model by floor. So, you would sort by building sections. On the other hand, if you are working on an estimate with a subcontractor who is interested only in the amount of gypsum board in the project, then you would be better off sorting by component types.

3. Select Quantities > Export to Excel (see Figure 2.20) to quickly export an estimate of items in the BIM to Microsoft Excel. This is not a linked file, and it is not a best-practice method of takeoff, but it is useful in quickly finding quantities of materials in the BIM.

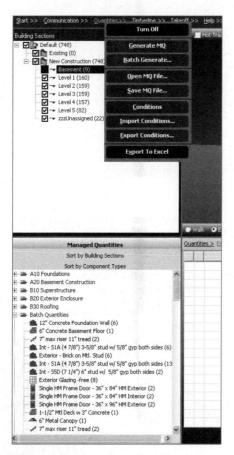

Figure 2.20 Exporting to Excel from Innovaya

4. In the Export Excel dialog box, specify the number of digits after the decimal (usually two) as well as the options to include the MQ name and MQ color. Then click OK.

This opens an Excel workbook (see Figure 2.21) that shows you square footage, width, height, level association, assembly codes, counts, and so forth. Although this is a good way to quickly get an idea of quantities from the model, any updated information will need to be sent through the entire process again. Costs still need to be assigned to these units. Typically this type of estimate is most useful as a way of getting an idea of cost and scope more quickly than On-Screen Takeoff or a digitizer during the concept or business development stage.

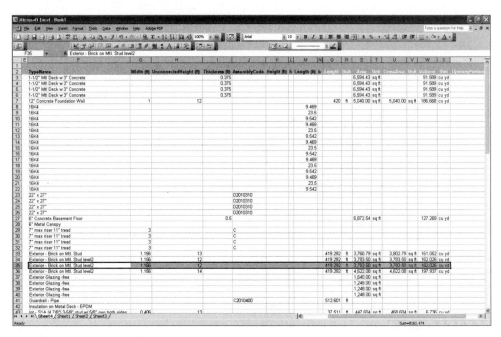

Figure 2.21 Exported Innovaya BIM takeoff in Excel

Innovaya Functionality

Innovaya can work with an MC2 estimate. This is useful to those who are already more familiar with the MC2 recipe format. By manipulating schedules and quantity tables in Revit, you can understand the amounts of materials in a given project. A company using Excel can format a standard set of schedules to copy into a Revit model and then export using the File > Export > Schedule command with the relevant schedule in the active window. This exports a delimited text file that can be opened in Excel.

Timberline

Next, link your model to a Timberline estimate, which is linked to a database.

Timberline estimating, from Sage Software, is a large database geared specifically toward estimators. The basic functionality of Timberline is to enable users to automate their estimating processes. This is achieved through the use of a database that houses assembly and model cost information. To clarify, these are not BIM assemblies but are similar in concept, because they are bundles of line items for a particular building component. These bundles of cost data include information such as material, cost, equipment, labor rates, and so on.

By compiling the data into a single item assembly, you can assign a cost to a metal stud wall, for example, associating multiple line items to the wall so you can use that item assembly in the future without having to create the assembly again. The assembly relates to the BIM in its linking to Innovaya by querying the components in the model. These are yes or no questions related to assembly costs. For example, after you link a model component with an assembly cost, the software asks, "Does the wall have blocking?" If you answer yes to this question, the estimate will enter the line item cost from the database to the current estimate. If you answer no, then it is not included. Because both Timberline and Revit are based on using assemblies to streamline their processes, it is a logical way of connecting estimate assemblies to model assemblies.

Creating a New Estimate in Timberline Estimating
1. Open Timberline Estimating.
2. Create and save a new estimate.
3. Close Timberline Estimating.
4. Proceed to the "Opening an Existing Estimate" steps.

Creating a New Estimate in Innovaya Visual Estimating
1. Select Timberline > New Estimate.
2. Proceed to the "Opening an Existing Estimate" steps.

Let's Talk Database

A preassembled database is available through Innovaya that contains a huge number of assembly mappings. This can be a great tool for new Timberline users or for users who self-perform a large portion of their work. If a large database is already in place in Timberline, experienced users can explore using existing assemblies and relevant cost histories, rather than starting over, which can be a large undertaking.

Opening an Existing Estimate

1. In Innovaya Visual Estimating, select Timberline > Open Estimate (see Figure 2.22). Specify the estimate you just created, and click Open.

   At the top of the window you should see two additional items with parentheses around them. The first is the database filename the estimate is linked to, and the second is the estimate name. Their presence here indicates that the estimate has loaded correctly.

2. To begin linking items, ensure that your MQ file is open and loaded into the Innovaya workspace.

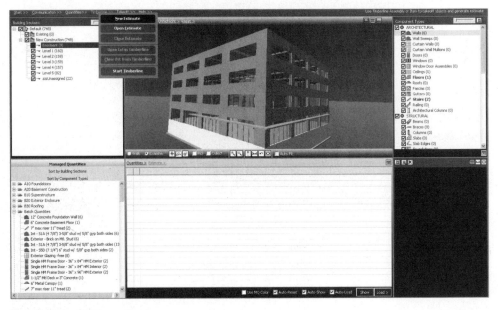

Figure 2.22 Creating a new estimate

Next, begin assigning costs. The Takeoff menu offers multiple ways to take off components from a BIM.

The first option is Assemblies/Items. Although your database might be configured a different way, the default database shows the two methods of taking off assemblies—as MasterFormat divisions under the Items category or as Assemblies items. Other options such as Quick Takeoff and One-Time Takeoff are not linked to any model components. The way Innovaya is set up allows users to assign cost line items to the estimate that aren't necessarily linked to a component. These items can be one-time takeoff items such as Formwork Set Up and Tear Down, Site Electrical Set Up, and the like. As in any construction estimate, there are costs associated with work that might not or cannot be modeled. There will probably never be a model the construction manager receives from the architect that includes all the form work, rebar layouts, and site excavation.

> **Note:** Hint: Try to limit one-time assembly takeoffs. Link costs to assemblies as much as possible; this will make reloading and updating much easier.

Assigning Costs

1. Click Sort by Component Types in the Managed Quantities pane, as shown in Figure 2.23.

Figure 2.23 Using Sort by Component Type to identify a BIM element

2. Expand the Level 1 listing and then the Walls category.

3. Click the first wall listed, 12² Concrete Foundation Wall.

   The wall becomes visible in the Detailed 3D Object Viewer pane at the lower right of the main window. The wall is also highlighted in the Dynamic 3D Model Viewer pane at the top center of the main window. Both of these views help clarify what component of the building you are estimating.

4. Select Takeoff > Assembly Takeoff. The Assembly Takeoff dialog box opens, as shown in Figure 2.24.

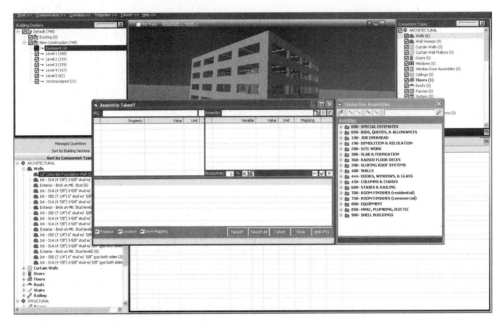

Figure 2.24 Innovaya's Assembly Takeoff dialog box

5. Now drag and drop the 12″ Concrete Foundation Wall item from the Managed Quantities pane to the MQ field at the upper left of the Assembly Takeoff dialog box (see Figure 2.25).

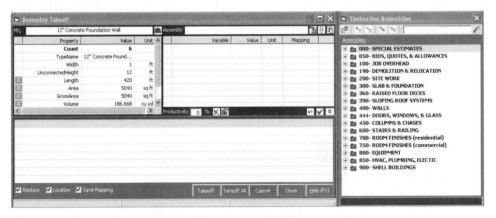

Figure 2.25 Dragging and dropping quantities into the Assembly Takeoff dialog box

6. In the Timberline Assemblies window, assign a cost link to the 12″ Concrete Foundation Wall by clicking the 400 Walls category to expand it and then dragging and dropping the Concrete Wall category on top of the blank Assembly pane (see Figure 2.26).

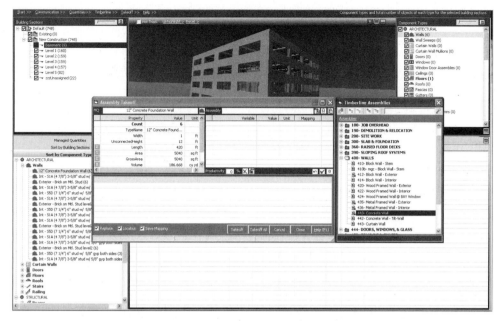

Figure 2.26 Dragging and dropping Timberline assemblies to the Assembly Takeoff dialog box

The Assembly pane shows values for several variables, beginning with the quantity. These values are tied to all cost assemblies. In the example shown, the selected wall from the Revit model (MQ pane) ties to the Timberline cost assembly (400 Concrete Wall) on the right.

7. Drag and drop the length and height from the MQ pane onto the corresponding variables in the Assembly pane to create a takeoff (see Figure 2.27). This specifies how you want to take off this wall; in this example, you are using use square footage and leaving the rest of the variables at their default values. For components such as doors, you might just drag and drop the count value.

Click Check Mark to Add a Pass

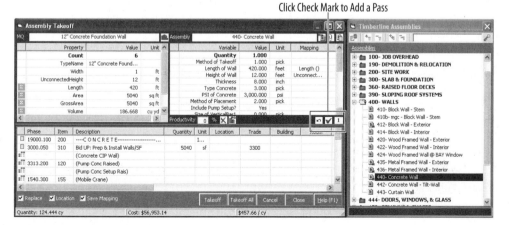

Figure 2.27 Assembly pane after takeoff

8. At the lower right of the Assembly pane is the Pass box; it's the one with a green check mark in it with 0 showing to its right. Click the green check mark to increase the number to 1. This field specifies how many times to add the assembly to the model. Although the majority of the time you will add just one pass to the model, some unique conditions might require multiple passes or counts to be assigned to an estimate.

9. Ensure that the Replace check box at the lower left of the Assembly pane is selected. This keeps you from accidentally creating multiple line items for the same wall.

10. Click the Takeoff button at the bottom of the Assembly Takeoff dialog box to add the wall to the estimate.

The concrete wall has now been added to the Timberline estimate and should be visible in the Estimate pane (see Figure 2.28). In the Managed Quantities pane, a yellow dollar sign appears to the left of the wall you just added to the estimate. This indicates that the component has been added to the estimate.

Yellow Dollar Sign

Figure 2.28 Wall added to estimate

Repeat the previous steps for the Int-S1A (4 ⅞″) wall, using the 436-Metal Framed Wall Interior Timberline assembly. Add a pass, and then click the Takeoff button (see Figure 2.29).

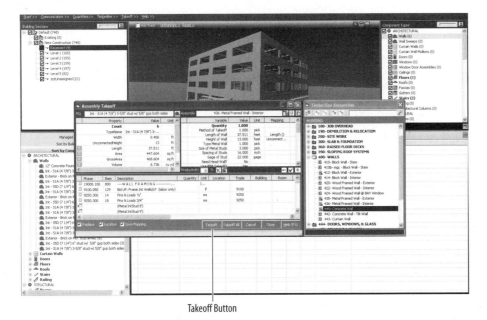

Takeoff Button

Figure 2.29 Using the Takeoff tool

Using the Auto-Takeoff Function

1. Select Takeoff > Auto-Takeoff (see Figure 2.30).

The Auto-Takeoff function assigns the same cost models to all model components with the same name.

Auto-Takeoff Button

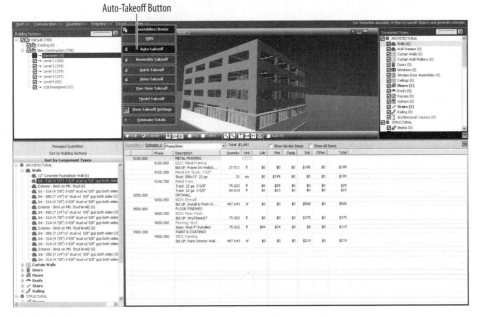

Figure 2.30 Auto-Takeoff on the Takeoff menu

2. Accept the default settings in the Auto-Takeoff window, and click the Go button.

The Object Property List shows that a cost has been assigned to all the Int-S1A (4 ⅞″) wall components in the model. The software keeps track of the model mapping as you continue to take off assemblies; after a model component has been dragged in, it automatically populates the assembly pane. Remember that an item is not added to the estimate until you add a pass and click the Takeoff button. This allows you to adjust for unique conditions, such as when one component has the same name as another assembly but needs to be altered to meet special conditions. For example, a soffit of the same wall type as Int-S1A (4 ⅞″) might have a unique cost because of its location or shape.

3. Continue assigning costs and mapping components to the estimate for the walls, roofs, and floors.

Doors are taken off as quantity-related items instead of square foot calculations.

Taking Off Doors

1. Drag and drop the Single HM Frame Door 36″ × 84″ onto the MQ field.

2. Drag and drop the 448-Opening-Doors onto the Assembly field.

3. Drag and drop the Count to the Quantity field that contains a 1 (see Figure 2.31).

Figure 2.31 Dragging and dropping quantities into the Assembly Takeoff window

This specifies that you want to take the item off by count. You can specify the height and width of the opening, the type of door, and the hardware later, but the primary means of taking off this element will be the quantity.

4. To finish taking off the door, specify the type of door and height and width of the opening, at a minimum, and then add a pass and click the Takeoff button.

5. Save both the INV file and the MQ file for use later in the book.

Site Coordination

Site coordination begins at schematic design and carries through the construction phase, as shown in Figure 2.32.

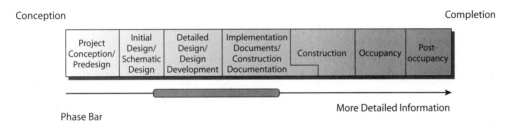

Figure 2.32 Site coordination timing

Site coordination is important for a construction manager, especially when dealing with dense urban environments or challenging sites. BIM offers tools such as perspective views and walk-through videos to show areas for crews to avoid during certain stages, check crane swings, show vehicular accessibility, and promote safety on site for workers, material hoists, equipment, and scaffolding. These views are called *site coordination plans* and can be constructed using a path of Revit into Google SketchUp or Revit into NavisWorks. Chapter 3 outlines how to create sequencing videos using NavisWorks. The following sections explain how to export a Revit model into Google SketchUp.

Starting the Site Coordination Plan in SketchUp

Google SketchUp (http://sketchup.google.com) is a free, intuitive program that is useful during the schematic and conceptual phase of projects for designers who want to understand the tectonics, scale, and massing of a design. You may want to purchase Google SketchUp Pro, which has added functionality and makes file importing and exporting easier.

Designers and architects often make SketchUp models before starting a Revit model, because of either their familiarity with the program or its ease of use. These SketchUp models can be shared and imported into the Revit model, just as a Revit model can be imported into SketchUp. Unfortunately, when you take the model from

Revit into SketchUp, you lose whatever intelligence the model might have had, and you retain only the 3D massing characteristics. That is why SketchUp is most useful toward the beginning of a project, when the concept is more of a priority than exact detail.

Exporting from Revit to SketchUp

1. Start Revit Architecture.
2. Navigate to your CD drive, and open example-core-shell.rvt; or download it from the book's companion web page, www.sybex.com/go/bimandconstruction.
3. Switch to 3D view in the Project Browser.
4. Select File > Export > CAD Formats (see Figure 2.33). The Export CAD dialog box opens.
5. Specify the type of file you want to export your 3D model as. Choose AutoCAD.
6. In the Export CAD Formats dialog box, click Options.
7. Select ACIS Solids as the export type. The default configuration of the export is a polymesh. This format makes it difficult to paint surfaces in some programs. The ACIS Solids format allows you to paint in SketchUp on single planar faces at once as opposed to painting onto multiple triangulated surfaces, such as in a polymesh file.

 Note: The advantage to exporting polymesh is for complex model geometries. This format is used in MAX, Maya, and Rhino for editing vertices.

8. Click Save to export the file and keep the file nomenclature the same.

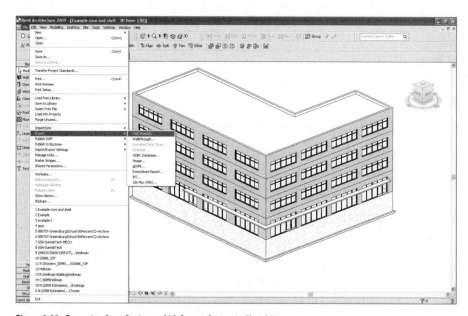

Figure 2.33 Exporting from Revit to a CAD format for use in SketchUp

Importing into SketchUp and Exploding the Model

1. Launch Google SketchUp.

2. In SketchUp, select File > Import.

3. In the Files of Type field, choose ACAD files (*.dwg, .*dxf).

4. Navigate to the CAD file you just exported, and, using the default settings, click Open.

5. After the model loads, click the selection tool (the arrow at the upper left of the SketchUp interface), and click the model. It should highlight in blue.

6. Right-click, and choose Explode (see Figure 2.34). This breaks the model into its original components.

The SketchUp interface is different from the Revit interface, but it is simpler and lacks intelligence to any associated objects. It does tell you the layer of the object and the name of the object, though. SketchUp is a valuable tool from a visualization and an ease-of-use standpoint.

Note: *Google SketchUp for Dummies* by Aidan Chopra (For Dummies, 2007) does a great job of explaining the ins and outs of SketchUp and the graphic interface and is extremely useful if you use SketchUp regularly.

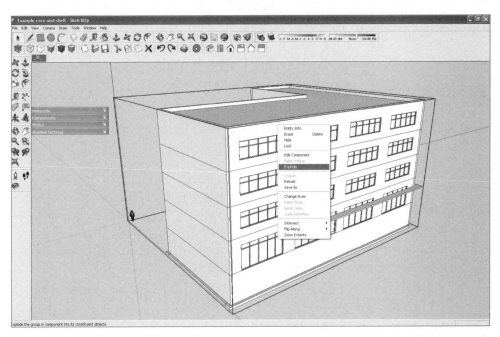

Figure 2.34 Exploding the model in SketchUp

Painting the Model in SketchUp

1. Click the Materials toolbar to expand it. (Clicking it again collapses it.) If the Materials toolbar isn't loaded in the viewer, activate it by selecting it in the Windows menu.

2. You can now begin painting the model by clicking a material in the toolbar and then clicking the model component you want to assign it to (see Figure 2.35).

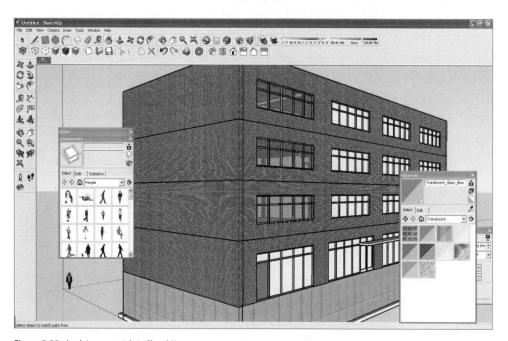

Figure 2.35 Applying materials in SketchUp

Inserting the Site

After you have rendered the building as desired, use Google Earth (http://earth .google.com/) to import the site into the model:

- The base version of Google Earth is available as a free download; it does a nice job of approximating terrain and scale, enabling you to visually place a building on a specific site. Google Earth Pro and Google Earth Plus are also available. Google Earth Pro is geared toward users who want to create high-resolution videos and presentations and use online collaboration tools.

- Google Earth Plus is focused on mapping, GPS, and terrain and civil information. Google Earth Plus offers real-time GPS tracking. Chapter 7 discusses asset management using RFID tags and GPS locators with Google Earth Plus.

  You can familiarize yourself with the Google Earth user interface at http://earth.google.com/intl/en/userguide/v4/. (See Figure 2.36.)

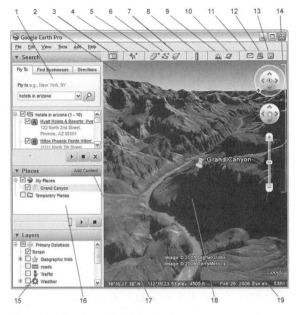

Figure 2.36 Overview of Google Earth

Importing a Site with Google Earth

1. Enter an address in the Search field. For this exercise, enter **West Watkins St & South 11th Ave Phoenix, AZ** (see Figure 2.37). Often when a building site doesn't have a formal address, the only information known is the intersection at the new building site.

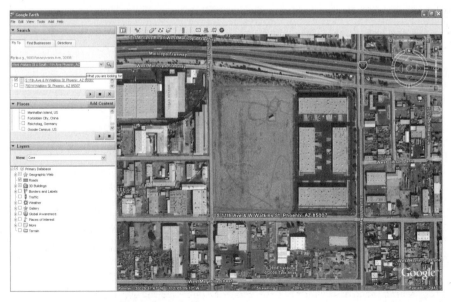

Figure 2.37 Inputting an address in Google Earth

2. Zoom into the area you want to use as the site, because what is visible in the window is what will be used as a reference for the site in Google SketchUp.

3. Toggle back over to Google SketchUp without closing Google Earth.

4. In SketchUp, move to the top or plan view.

5. Click the Get Current View button; this is the world icon with a yellow arrow over the top of it (see Figure 2.38). It imports the current Google Earth view into the SketchUp model.

   The result is a Google Earth terrain image and the SketchUp model.

6. Place the model on the site by moving the model on top of the site image and positioning it. This might involve rotating and using the move commands in SketchUp.

   After your building is positioned, it should look like Figure 2.39.

7. Click the Toggle Terrain button (see Figure 2.40) to view the slope of the terrain.

Figure 2.38 Get Current View button in SketchUp

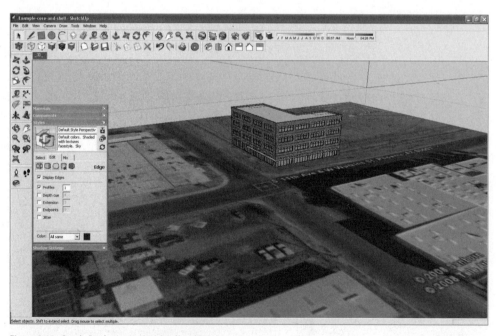

Figure 2.39 Google Earth terrain placed in SketchUp

Figure 2.40 Toggle the view between a flat photo of site and a 3D representation of terrain with the Toggle Terrain button.

In this example, the site is relatively flat, so it looks like there won't be any issues with the first-floor storefront glazing wrapping around the building at the same level. However, if you put this model in San Francisco at the intersection of Montclair Terrace and Lombard Street, you get quite a different story (see Figure 2.41)!

Figure 2.41 Building placed on the steeply sloped Lombard Street in San Francisco

The last portion of this tutorial involves drawing additional information on the site coordination plan. These elements can include directional arrows, 3D text, staging areas, worker parking, and a host of other information (see Figure 2.42). SketchUp lets you quickly input general information to communicate how a construction manager wants to operate on the site.

Figure 2.42 An example of a site coordination plan

Using the Line and the Push and Pull tools, you can create virtually any shape or outline you want to use to communicate your plan more effectively.

SketchUp also exports into Google Earth. Importing the SketchUp model into Google Earth lets the user get an idea of site context, surrounding building scale, access to resources, distance calculations, and adjacent infrastructure.

Exporting a Site Coordination Plan to Google Earth

1. Click the Place Model button in SketchUp (see Figure 2.43). This inserts the SketchUp site coordination plan into Google Earth temporarily (see Figure 2.44) so you can use it to analyze a particular site.

Using Google Earth and Google SketchUp for site analysis lets you gather site information without a completed civil survey to understand grading, property boundaries, staging areas, and the best means of loading and moving equipment. Overall, the Google software is a powerful resource when doing preliminary site analysis, and it is relatively easy to use and understand. These tools provide an open platform for online collaboration, letting you add, edit, and share 3D information about the site.

Figure 2.43 Clicking the Place Model button exports the SketchUp site coordination plan to Google Earth.

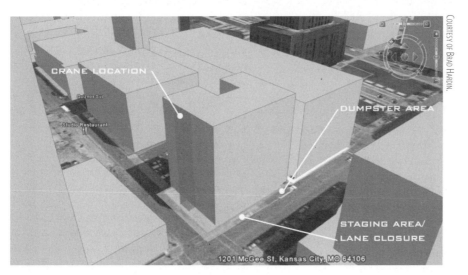

Figure 2.44 A model placed into Google Earth, showing a 3D view

Conclusion

Using BIM and BIM-aware tools such as those described in this chapter can give you an early understanding of what the scale, site, and building conditions are and what the cost of a project is. Using BIM, and the technologies that support its varied uses, helps inform the project team and streamline processes. You can use site analysis tools to understand multiple sites and communicate potential advantages and drawbacks of each without leaving your desk. As a design model changes throughout the course of a project, you can use the BIM as a tool to coordinate takeoff and costs. Using this means of estimating creates a proactive design-to-budget approach with the architect and engineering team, as opposed to dealing with a reactive response to value engineering conditions later. Using BIM involves the use of both analysis tools and modeling tools, and both are critical to a systemic BIM process. Understanding that there is a difference between the two tools will clarify future tasks and programs. This book shows which tools (and identifies similar tools) are applicable to each task, a solution to how it can be accomplished, and the ability to quantify the time and effort it takes to accomplish these tasks for future reference.

BIM and Construction

This chapter explores BIM during the beginning phase of a construction project, outlining what BIM can mean to a construction project and defining the amount of work associated with three topics: scheduling, constructability, and multiple trade coordination. This chapter also includes step-by-step tutorials for each of these tasks to show how BIM can provide tools along the project management path to increase a project's efficiency. Specifically, this chapter covers the following topics:

3

In this chapter

Scheduling

Constructability

Trade coordination

Many tasks can be associated with construction and project management, and scheduling, constructability, and trade coordination represent a good cross section of how BIM may be used. I will identify how these tasks can be accomplished in concert with the architectural and engineering teams. I will begin by explaining some of the fundamental deliverables of a construction manager currently and how BIM can enable better coordination in a "typical" construction project.

To some companies, using BIM during the construction phase may mean implementing a new process and refining operational tasks. Although some companies use BIM throughout the entire course of a project, many others either stop using it in the preconstruction phase or use various bits and pieces of a BIM process to help them better coordinate a project. Although there's no right or wrong way to use BIM, the most important question is, how can you improve the way you practice construction management? As projects progress, it is easy to slip back into the same old way of doing things, and in doing this, there is very little chance for future growth, either in technology or in efficiency. Gradually adopting BIM initiatives is the best way to change existing practices until it becomes habit. As projects become more advanced, complex, and difficult, the technology used in these projects will also advance along with them. More exciting, the construction industry is driving these technologies further then ever before. Entrepreneurs, software companies, and tech-savvy professionals are developing BIM tools rapidly to meet the rising demands of the industry.

Scheduling

Before the first shovel of dirt moves, the construction manager must do a significant amount of work. Arguably, the most important task is preparing the schedule. The schedule is one of the driving factors in the success of any project and is a critical component to all team members. At the onset of a project (and often before that), a member of the construction management team creates the schedule. This schedule typically reflects experience with construction timing; material lead times, weather, crew, and equipment concerns. The importance of a schedule in regard to BIM is to better inform the team and track progress from the beginning to the end of a project (see Figure 3.1). So, how does BIM improve schedule management? How can you use BIM to increase visualization and schedule accuracy?

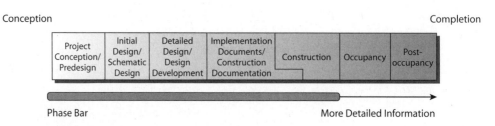

Figure 3.1 Updating a BIM schedule is a continuous task through a project.

A construction schedule is a sophisticated chart or table showing tasks and the times required to accomplish them (Figure 3.2). Although there is an implied correspondence between a task and a building component, there really is no direct link between the CAD drawings, the specifications, and the construction schedule. As the design progresses, the construction manager reviews the updated drawings to identify changes in scope, as well as the addition of design elements, and then updates the schedule to reflect these design changes. The refinement of the schedule relies on the accuracy of the construction manager to review the new design documents each time and judge the projected availability for additional equipment, material quantities, and so on. The schedule, and the subsequent revisions, is one of the more time-intensive aspects of a project, and the members of the team rely on its accuracy to deliver a project to the owner on time. Therefore, any increase in efficiency and schedule accuracy can do two things. First, it would provide the construction manager with more time to further coordinate other tasks. Second, it mitigates many of the issues associated with schedule misinterpretation through enhanced visualization by linking the schedule to the virtual construction. This is the ultimate goal of BIM—to increase efficiency, communication, and collaboration.

Note: I mean *efficiency* here not only in terms of time but also in terms of costs, accuracy, and thoroughness.

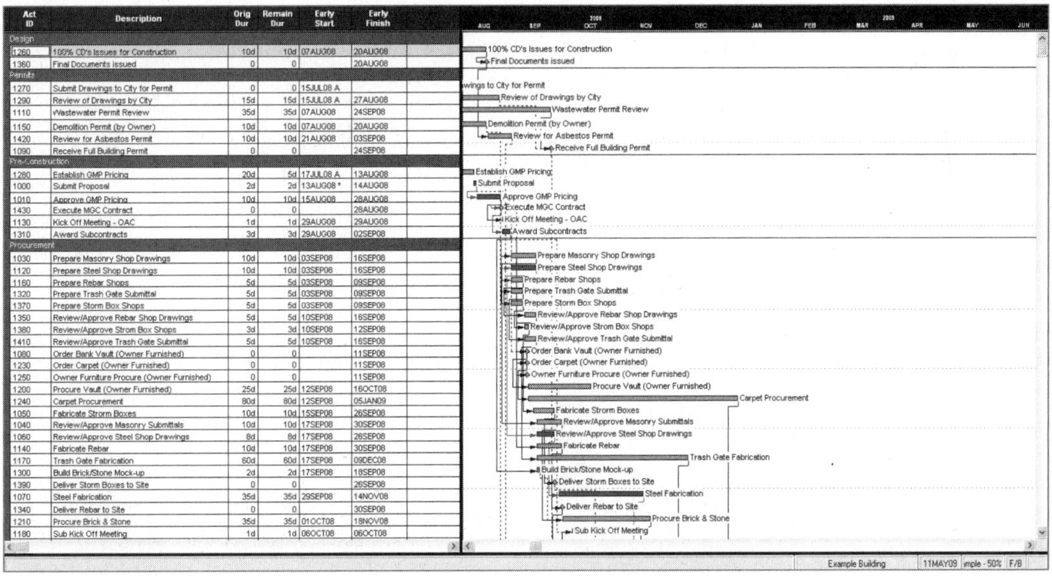

Act ID	Description	Orig Dur	Remain Dur	Early Start	Early Finish
Design					
1260	100% CD's Issues for Construction	10d	10d	07AUG08	20AUG08
1360	Final Documents issued	0	0		20AUG08
Permits					
1270	Submit Drawings to City for Permit	0	0	15JUL08 A	
1290	Review of Drawings by City	15d	15d	15JUL08 A	27AUG08
1110	Wastewater Permit Review	35d	35d	07AUG08	24SEP08
1150	Demolition Permit (by Owner)	10d	10d	07AUG08	20AUG08
1420	Review for Asbestos Permit	10d	10d	21AUG08	03SEP08
1090	Receive Full Building Permit	0	0		24SEP08
Pre-Construction					
1280	Establish GMP Pricing	20d	5d	17JUL08 A	13AUG08
1000	Submit Proposal	2d	2d	13AUG08 *	14AUG08
1010	Approve GMP Pricing	10d	10d	15AUG08	28AUG08
1430	Execute MGC Contract	0	0		28AUG08
1130	Kick Off Meeting - OAC	1d	1d	29AUG08	29AUG08
1310	Award Subcontracts	3d	3d	29AUG08	02SEP08
Procurement					
1030	Prepare Masonry Shop Drawings	10d	10d	03SEP08	16SEP08
1120	Prepare Steel Shop Drawings	10d	10d	03SEP08	16SEP08
1160	Prepare Rebar Shops	5d	5d	03SEP08	09SEP08
1320	Prepare Trash Gate Submittal	5d	5d	03SEP08	09SEP08
1370	Prepare Storm Box Shops	5d	5d	03SEP08	09SEP08
1350	Review/Approve Rebar Shop Drawings	5d	5d	10SEP08	16SEP08
1380	Review/Approve Strom Box Shops	3d	3d	10SEP08	12SEP08
1410	Review/Approve Trash Gate Submittal	5d	5d	10SEP08	16SEP08
1080	Order Bank Vault (Owner Furnished)	0	0		11SEP08
1230	Order Carpet (Owner Furnished)	0	0		11SEP08
1250	Owner Furniture Procure (Owner Furnished)	0	0		11SEP08
1200	Procure Vault (Owner Furnished)	25d	25d	12SEP08	16OCT08
1240	Carpet Procurement	80d	80d	12SEP08	05JAN09
1050	Fabricate Strorm Boxes	10d	10d	15SEP08	26SEP08
1040	Review/Approve Masonry Submittals	10d	10d	17SEP08	30SEP08
1060	Review/Approve Steel Shop Drawings	8d	8d	17SEP08	26SEP08
1140	Fabricate Rebar	10d	10d	17SEP08	30SEP08
1170	Trash Gate Fabrication	60d	60d	17SEP08	09DEC08
1300	Build Brick/Stone Mock-up	2d	2d	17SEP08	18SEP08
1390	Deliver Storm Boxes to Site	0	0		26SEP08
1070	Steel Fabrication	35d	35d	29SEP08	14NOV08
1340	Deliver Rebar to Site	0	0		30SEP08
1210	Procure Brick & Stone	35d	35d	01OCT08	18NOV08
1180	Sub Kick Off Meeting	1d	1d	06OCT08	06OCT08

Figure 3.2 A schedule is a series of complex, overlapping tasks to ensure successful project delivery.

In the tutorials later in the chapter, you will simulate real processes by using an architect's example model to generate a BIM *sequencing animation*, which is the link between the design model to the construction manager's schedule and is an extremely valuable tool. As more model components are added and as the schedule changes, linking eases the work associated with updates while still providing a robust resource.

A *scheduling animation* shows in 3D the building being built from start to finish; it helps communicate completion dates to owners, gives subcontractors and tradespeople a better understanding of the scope and timing of their work, and helps field personnel verify the project is on track. To use BIM for scheduling, the model-sharing language discussed in Chapter 2 must in place. If language hasn't been established, there may be some challenges for the contractor to receive the architect's schematic or design development–level BIM file. Although there is not a huge need to go into great detail about sharing the model during this phase, contractors should state their intentions for using the model in the IE responsibility plan. Additionally, the construction manager must also understand that a design development–level BIM is by no means a completed BIM model. In a BIM process, it is helpful to establish an understanding that model sharing is critical to accomplishing more integration, especially if the construction manager is to advise the design team through the preconstruction phases. After the model request has been made and you've received the model from the architect, you can begin creating a scheduling animation.

The advantage to beginning with the architect's model is threefold:

- During the design phase, it tells you what components have been modeled and to what level of detail. Users who are experienced working with building information modeling can tell a well-constructed model from one that needs work. This almost becomes second nature, just as a construction manager can spot well-coordinated print documents today.

- There is a cost savings in not allotting additional resources to remodeling a structure very much in its infancy. Creating a secondary construction BIM model is typically unnecessary in a project, especially during the early design phases of a project when it is a waste of resources.

- The construction manager can identify additional elements, design updates, and program changes for reference, as well as begin separate layering construction model information, discussed in more detail later in this chapter.

Although these are all benefits, the main benefit to beginning with the architect's model is that you're using the product developed by the design team, which is a best practice. Many times, the response to a design development–level BIM model by a contractor is that it is not complete enough and that "I am going to have to create my own model instead." A model typically changes many times prior to creating the construction documentation. It is better to use a single model and inform the architect of any big issues associated with their model as opposed to putting it by the wayside. Using the architect's model helps to identify new items and scope as well as coordinate owner-driven design and program shifts; in other words, coordinating once instead of twice.

Scheduling Software

Scheduling software, such as Primavera (www.primavera.com) or Microsoft Project (http://office.microsoft.com/project), keeps track of the work breakdown structure (WBS) and critical path dependencies between overlapping tasks to create complex timelines, which can be displayed in a variety of standard formats. The software can be used for planning as well as for tracking projects once they are underway. Schedules are constantly updated, and the software helps update the project's schedule.

Both Microsoft Project and Primavera systems are compatible with Navisworks TimeLiner. So is any other scheduling software that can produce an MPX or a Primavera version 5 file. The following tutorial uses Primavera to demonstrate how to link a static schedule to a BIM schedule. The power of most scheduling software is its ability to easily overlap, link, and create very complex schedules with large amounts of tasks tied to a timeline. Updating these schedules can be a constant source of work that is required to define the progress of a project. These scheduling programs and others simplify the task of creating these complex schedules and are commonplace in the industry.

Starting with an existing Primavera schedule (one is available on the book's companion web page, www.sybex.com/go/bimandconstruction), you can link a simple schedule during the conceptual stages of a project and add detail later.

> **Note:** Remember when working from Primavera that future revisions and changes supersede the old schedule. Always archive your old schedule, and save over the old schedule with the same filename.

Exporting a Primavera Schedule for Use in Navisworks

1. Download the file Example-50% DD from the book's companion web page.
2. Start Primavera SureTrak, and open the downloaded file.
3. Select File > Save As.
4. Select the MPX file type (Figure 3.3), and save the file.

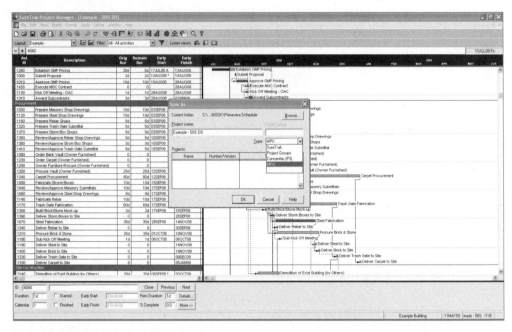

Figure 3.3 Saving the schedule as an MPX file

Next you will import the saved MPX file into Navisworks to begin linking model elements to the schedule.

Navisworks Collaboration Software

Autodesk Navisworks (www.autodesk.com/navisworks) is a powerful tool for construction managers using BIM. Navisworks is collaboration software that allows a design team to share, combine, review, and correct a BIM model and 3D files using a 3D viewer. Navisworks can open multiple 3D files and combine them in a single workspace. Navisworks or similar software, such as Solibri Model Checker (www.solibri.com), can provide functional insight into the growing variety of industry software systems.

Many subcontractors, such as fabrication and sheet metal shops, may already be using 3D modeling software that generates information you can integrate into a BIM workflow. The typical shop deliverable file is 2D sheet drawing that does not reflect the 3D design, because the 3D modeling information may not have been requested.

Fabricating from a 3D Model

Cates Sheet Metal, a ductwork manufacturer in the Midwest, has been creating 3D models for some time and shipping them to computer numerical control (CNC) machines to laser cut the sheet metal and fold them into the correctly sized components (Figure 3.4).

IMAGE COURTESY OF CATES SHEET METAL.

Figure 3.4 Using a plasma cutting CNC machine

This company is similar to others in that they have modeled duct runs, connections, and other components in 3D for some time. However, as the standard requested deliverable has been 2D sheet drawings derived from the 3D model, many architects and contractors don't know to request the 3D files as well for shop drawing review.

Although this example uses sheet metal, many other fabrication shops coordinate in 3D, such as structural steel, casework, precast concrete, fire protection, piping, and other specialty fabrications. Many machines on the market today use 3D models to fabricate their components to pinpoint accuracy based on 3D information and coordinates.

Navisworks is not modeling software, but rather analysis software. In the tutorial to follow, it allows models to be compiled and linked to a schedule to create a schedule animation.

Composite Modeling

Composite modeling is a modeling compilation strategy that combines the available 3D information into a single shared file. Composite modeling is not necessarily the ability to house all team members in the same office, developing the same model, using the same software. Although some companies are capable of this type of model development with architects, engineers, and contractors in house, it is unusual. Some owners who have a fast-track project using BIM find that one way to rapidly advance a project is to have a *BIM pit* or *BIM huddle*, in which all the members of the team, even those from different companies, have their office in one location, where they work together to model and virtually construct the proposed structure. However, it is more common that the design team uses a singular composite model. A composite model is a series of 3D models that are created from the same or different pieces of software and that can be compiled for analysis and advanced visualization. Arguably, the most robust tool in which these models are compiled and tested is Navisworks.

How Many Files!?!

One of my first BIM projects was for a medical facility. Part of the IE responsibilities was to combine the multiple 3D formats of the project and create a model clash detection report. Initially, I was tentative about this undertaking, because the project had been put on hold for some time and had just come to the forefront again with such vigor that the management team had to move fast. Our standard preliminary meetings were very compressed.

The owner desired a BIM deliverable project, and all parties involved knew that the project was to be finished in BIM; however, the discussion late on a Friday afternoon quickly turned to how to do it. The architect was using Revit Architecture, the structural engineer was using SDS2 modeling software, the mechanical subcontractor was using CAD-Duct, and the civil engineer was using AutoCAD Civil 3D. We anticipated using Navisworks for clash detection for the most part. However, using these other files to generate a clash report was new science to us. But we discovered that every 3D model that had been created could be compiled into the Navisworks model. Although this initial multiple-file undertaking went smoothly as we generated our clash detection report, we learned to double-check whether we could use the team's native formats or had to use exported versions prior to compiling the file in Navisworks.

This book shows how to use Navisworks to run a schedule animation, sequencing animation, and clash detection. Navis has other tools, but these three are the ones used most by construction managers. The greatest benefit to using Navisworks is the ability to combine many files of many different file types. Again, Navisworks is not a modeling program; rather, it links BIM and 3D files into a Navisworks format (NWD), which is often a more useable file type than the NWF format. Both files can be viewed using the free viewer for Navisworks files, called Navisworks Freedom. This viewer is useful for those who might want to look at conflicts or at the composite model overall but who don't want to purchase the full version or any licenses of Navisworks. Figure 3.5 shows the basic Navisworks interface.

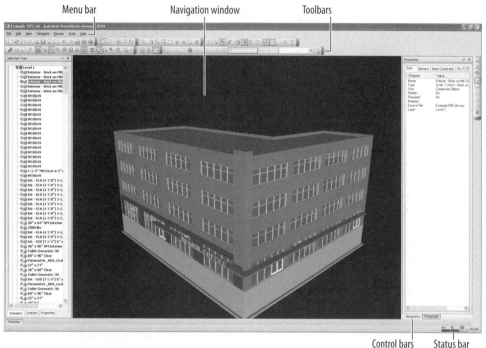

Figure 3.5 Navisworks user interface

As mentioned, the purpose of a scheduling animation is to show in 3D the building being built from start to finish. Ultimately, the quality of the animation is directly related to the quantity and accuracy of the model components. Keep in mind that it requires additional time to link more components to the schedule. Furthermore, the more complex the schedule, the longer it takes to link to more lower-level schedule elements. In a schedule animation, you can show the earthwork excavation, site demolition, pile driving, piers, excavation, forming, site utilities, crane erection, truck loading areas, staging and lay down areas, reinforcement and rebar, concrete foundation pour, structural steel erection, and so on. Almost any activity that occurs during construction can be modeled if represented by a virtual model component. With Navisworks, you can create detailed or simple animations using these 3D model components.

Exporting a Revit Architecture File for Use in Navisworks

1. Download `Example-50% DD.rvt` from the book's companion web page.

2. Launch Revit, and open the file.

3. Choose Tools > External Tools > Navisworks 2009 (Figure 3.6). The Export Scene As dialog box opens with the Navisworks NWC file type selected by default. Change the linear units to Feet and Inches (Figure 3.7).

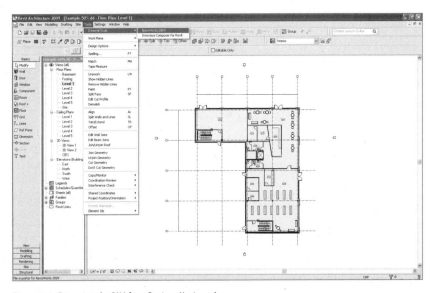

Figure 3.6 Exporting the BIM from Revit to Navisworks

Figure 3.7 Changing the export settings to feet and inches

4.   Click the Navisworks Settings button to open the Options Editor dialog box. The settings in this dialog box define how you want to export your file to Navisworks.

5.   In the tree at the left, expand the Interface branch, and select Display Units.

6.   Change the default setting from Meters to Feet and Inches.

7.   Select Snapping, and choose Snap to Vertex, Snap to Edge, and Snap Line to Vertex (Figure 3.8).

8.   Click OK to exit the Options Editor.

9.   Specify where you want the NWC file to be saved, and click Save.

Figure 3.8 Enabling snaps in Navisworks

Importing the Model into Navisworks

1.   Launch Navisworks Manage 2009.

2.   Choose File > Open, and navigate to the NWC file you just created.

3. Choose File > Save As.

4. Specify the NWD file type, and click Save (Figure 3.9).

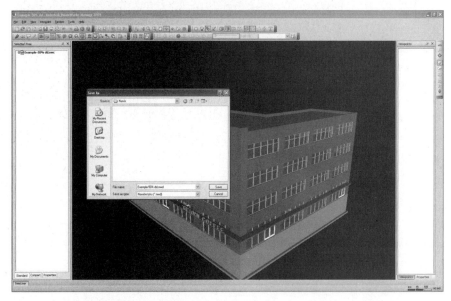

Figure 3.9 Saving the NWC file as an NWD file

Now that the model is saved in Navisworks, you need to import the Primavera schedule you exported earlier.

Importing the Schedule into Navisworks

1. Click the TimeLiner button on the toolbar (Figure 3.10). The TimeLiner window opens at the bottom of the screen.

2. Click the Links tab, and right-click in the blank table area to open a context menu.

Figure 3.10 Activating TimeLiner in Navisworks

3. Choose Add Link > Microsoft Project MPX (Figure 3.11) to link the MPX file you exported earlier, and click Open. This opens the Field Selector dialog box.

4. Select Text10 in the Unique ID Import Field drop-down menu (Figure 3.12), and click OK. This adds the link to the TimeLiner window.

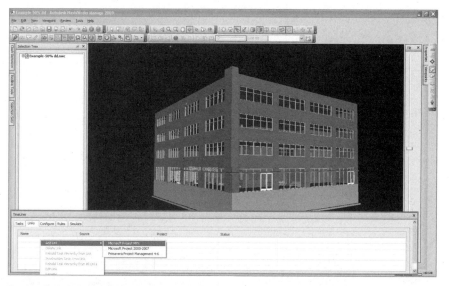

Figure 3.11 Linking the MPX file

Figure 3.12 Selecting the Text 10 Unique ID field

5. Right-click the new link, and choose Rebuild Task Hierarchy from Link on the context menu (Figure 3.13). This takes all the schedule line items and breaks them out into tasks within Navisworks.

6. Click the Tasks tab. All the line items in the schedule are now tasks, with start and end dates.

Figure 3.13 Rebuilding the task hierarchy from the link

Now you can begin linking tasks to model components. You can go about this in a couple of ways:

- Navisworks includes a search tool that allows the model components within Navis to be searched and grouped based on the name type.

- You can assign tasks to model components manually.

A model search is usually the easier way to link schedule items to the model. As similarly named items are added later, it finds these new elements with the same specified search parameters and links them automatically.

Linking Tasks to Model Components

1. Click the Find Items button in the toolbar (Figure 3.14).

Figure 3.14 The Find Items tool

2. Select the file Example-50% dd.nwc, and select the following values in the fields on the right, as shown in Figure 3.15, by clicking in the field and choosing from the drop-down list.

Field	Value
Category	Item
Property	Type
Condition	Contains
Value	footings

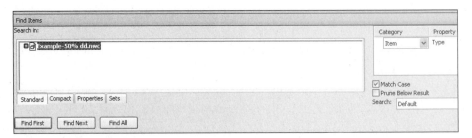

Figure 3.15 Setting the search parameters

3. Click the Find All button. In the 3D browser pane, all the footings in the model are highlighted.

You can create search sets for all the listed and available categories in Navisworks, which helps delineate one component from another more easily. In the next set of steps, you'll save this search set in Navisworks.

Search Sets vs. Selection Sets

Navisworks allows you to create two types of selection groups:

- A selection set groups components together that have been manually selected by the user, either through the 3D browser or through the model selection tree.
- A search set groups components based on search criteria.

The advantage to creating search sets as opposed to selection sets is that search sets allow the model to be updated and the model components to be more easily selected. In a selection set, the selection would have to be manually updated with each subsequent model update. So, it is usually a better practice to use search sets as opposed to selection sets.

Creating a Search Set in Navisworks

1. Using the previous search selection, click the Selection Sets tab on the left edge of the Navisworks window. When the Selection Sets pane opens, dock it by clicking the pushpin at the upper right.

2. Right-click in the open area, and select Save Current Search from the context menu.

3. Name this search set **footings**.

4. After you've created the search set, scroll down the task list in the TimeLiner window.

5. Right-click the task Footings and Foundations, and select Attach Search from the context menu (Figure 3.16). The updated status is reflected on the Footings and Foundations line in the listing, indicating the search has been linked to the task successfully.

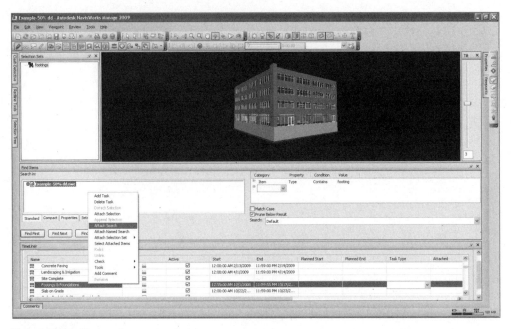

Figure 3.16 Attaching search sets to the schedule

6. Click the Task Type field in the same line, and select Construct from the drop-down list to indicate that these are construction, rather than demolition or temporary (such as shoring or formwork).

7. To verify that the simulation is being set up correctly, click the Simulate tab, and then click the Play button. The simulation should show the footings being constructed along the project timeline. If the animation is moving too quickly, click the Settings button, and specify the desired interval and playback duration (Figure 3.17).

Manually Assigning a Single Component to the Schedule

1. Select a foundation wall in the 3D browser.

2. Right-click the Footings and Foundations task again, and select Attach Selection from the context menu (Figure 3.18).

You can also assign model components to tasks using the selection tree. You'll do this in the following tutorial.

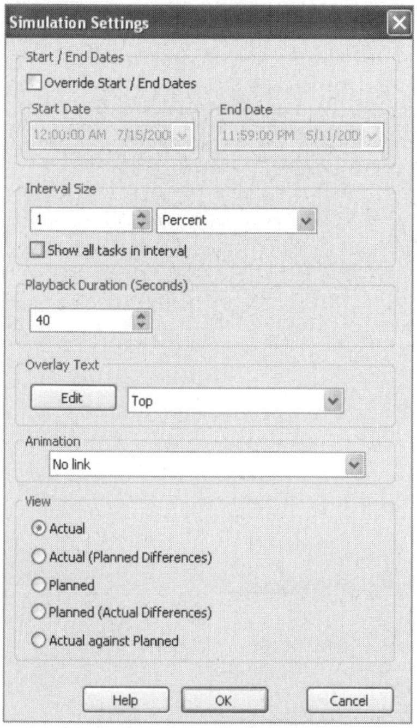

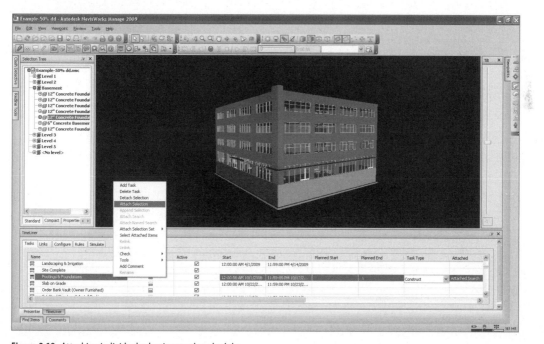

Figure 3.17 Editing animation settings

Figure 3.18 Attaching individual selections to the schedule

Assigning Components Through the Selection Tree

1. To open the model selection tree, click the Selection Tree icon in the toolbar (Figure 3.19). Dock the selection tree in the window by clicking the pushpin in the upper right.

Figure 3.19 Attaching through the selection tree

2. Select Basement Level. You can then select all the basement walls or just the related basement wall.

3. Right-click the Footings and Foundations task again, and choose Attach Selection from the context menu.

4. Continue to link the rest of the model to the schedule as desired.

Using this powerful tool in Navisworks, you can simulate a schedule in 3D to better communicate the order and construction of a structure. The animation can be exported as a rendered animation for business development purposes. Creating the animation takes effort the first time it is being developed; however, updates usually take much less time to create. You can use a scheduling animation in the field to indicate degree of completeness, which helps assign some visual basis of completion to contractors and subcontractors. In essence, BIM bridges the gap between model component and schedule and is an invaluable tool to the construction manager using BIM.

Constructability

Constructability, *means and methods*, and *project construction feasibility* all refer to the evaluation of whether the design can actually be built by a construction team and how it will be done. To a construction manager, constructability issues can affect many components of a project including costs, schedule, materials, and labor. To an architect, constructability means the ability for the design to be constructed as envisioned. To an engineer, constructability may mean that the actual construction meets the performance criteria set forth. In many contracts, the contractor is responsible for the means and methods of constructing a project, based on the *design intent* documentation. Constructability reviews typically begin during the design development phase of a project, as the design begins to mature, and lasts through the beginning of construction (Figure 3.20).

Constructability covers everything from logistics to concrete pour schedules to material procurement to crane erection and teardown, and so on, to determine just how a project is to be constructed and the site orchestrated. While the architect is envisioning the space of a floor, layout, flow, and life safety issues, the contractor is visualizing the logistics, including formwork, control joints, slab edge conditions, and pour sequences.

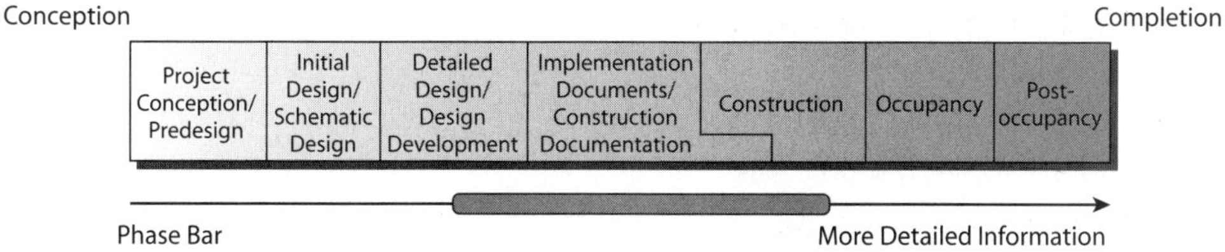

Conception   Completion

| Project Conception/ Predesign | Initial Design/ Schematic Design | Detailed Design/ Design Development | Implementation Documents/ Construction Documentation | Construction | Occupancy | Post-occupancy |

Phase Bar   More Detailed Information

Figure 3.20 Constructability reviews begin during design and last until construction.

The structure must be thought through from the unique perspectives of the design and construction professionals. The two are very much intertwined, and constructability issues are just as important as the design issues. This interdependence is integral to a BIM process. Working through both issues together can produce incredible results.

The current method for architects to use BIM to review constructability is still largely undefined. Some architects use BIM to create design intent documentation just as they have with CAD, while others have determined design intent is no longer a feasible solution when using building information modeling. The architect is not as limited by technology as before and has the ability now to create a very accurate virtual construction of a building. The word *accurate* is used in the context that although no set of documents printed or virtual can present an exact replica of what is to be constructed, BIM allows you to get closer than ever before. Therefore, today's architect has two paths to choose from: design intent documentation or virtual design/construction documentation.

Although the legal world still is determining the implications of a VDC (Virtual Design and Construction) model, the decision to move toward a more accurate modeling practice and even directly to VDC BIM has become popular for a number of reasons:

- First, an architect rarely regrets creating more accurate documentation, and as modeling proficiency increases, the ability to model quickly facilitates additional information in the process.

- Second, working toward a collaborative and interoperable system of using BIM is the future. Although some software limits the comprehensive capabilities of BIM because of a lack of interoperability, the future of BIM is moving toward a virtual, buildable model.

- Third, emulating actual construction in modeling becomes a useful tool for the construction manager building it. This is especially true when compared to typical CAD design intent documentation and is evident in construction managers creating their own BIM model for 2D projects. Unless an architect practices total prefabrication such as prefabricated housing or modular installations, a typical building involves a hybrid of on-site construction and off-site fabrication. To this end, architects and contractors are in a unique position to create some portion of the construction, which may feed directly to a fabrication model. For centuries, architects and builders have practiced with drawings and scale models representing their three-dimensional vision for what a building is to become. This is not because it is the best way to do so but because the technology available to the profession dictated the form of representation. With the advent of BIM, you can see a significant change in the ability to accurately represent the building as desired to be constructed.

Contractors are not waiting for architects to get on board either. In fact, in many projects where architects have chosen to use CAD, contractors now create a completely separate model based on an architect's 2D documentation to increase their own efficiencies and to better understand how the design intent representation is to be constructed. Because the majority of risk and responsibility lies typically with the contractor, progressive contractors are using all the tools available to make a job run smoothly and mitigate risk. However, the ability to create a separate model to a satisfactory level of detail can require a significant amount of resources. Because the number of construction companies in the United States that employ 500 employees or less far outnumber the companies that have 500 or more, personnel and financial resources can be limited. This means the ability for a contractor to perform the virtual construction of an entirely separate model might not be feasible for mid- to smaller-sized firms. Embracing a hybridized approach, where the architect and engineers have provided, at a minimum, base models gives the majority of contractors the ability to engage in a BIM process to some level without a significant out-of-pocket investment for each project. Unfortunately, if the project is in 2D, the ability to utilize any BIM intelligence is limited and involves re-creating a model derived from the CAD drawings, if the contractor still chooses to use BIM on a project.

Other resources, such as outsourcing, have risen in popularity. This process involves engaging a shop in another country such as India or China as a means of creating a model from 2D CAD documentation, usually because it costs less. Although this methodology assists a contractor in meeting BIM requirements, the model quality needs to be checked when sent from shops for errors in measurement, because of conversions and other inaccuracies, until a trust level is established with the resource.

In all of these scenarios, there is no longer an advantage, need, or excuse for design intent documentation as it currently exists when using BIM. Although this might seem severe, both owners and contractors alike realize there is little value in the 2D documents being produced and therefore have started using BIM heavily to meet their needs. In addition, creating design intent BIMs doesn't offer much to end users, such as facility managers and owners, who will still have to re-create and edit information to meet their needs. The architecture profession must realize the very real challenge in the way architects practice. The design community now has a very real opportunity to recapture responsibility lost by being more involved in creating a virtual construction model, as well as more say in the design. Conversely, architects and engineers may continue to relinquish even more responsibility to the contractor, which could present a number of issues to all professions involved.

Perhaps the highest process hurdle to jump over is the idea that BIM equals what is to be built. As discussed earlier, a model doesn't need to three-dimensionally represent every last doorknob and hinge, because that is neither productive nor efficient; rather, the model contains enough information that it can be constructed. This presents architects with an opportunity that many design-build companies have already capitalized on: designer and contractor collaboration. Creating a VDC BIM results in better delivered projects as it engages both the designer and the builder to coordinate the information prior to any construction. Using a model planning strategy (such as the IE and Model Coordination plan) can limit modeling times, increase document quality, and improve field collaboration when construction begins.

It has been my experience that creating detailed and construction-ready BIM models is a more heavily involved process than creating CAD documentation but that it doesn't necessarily use more resources. For example, it might take an architectural team of six to eight team members to document the example BIM model in this chapter. However, this same project might take three to five team members if using BIM software. The time for creating the BIM model might be the same, but because the team size is reduced, it results in some savings. This savings in general terms gives architecture firms the ability to take the BIM model to the next level through additional collaboration and modeling time. In essence, the overall fee and overall profit may not increase, but the efficiencies within the team increase, and the process produces better-quality documentation (Figure 3.21).

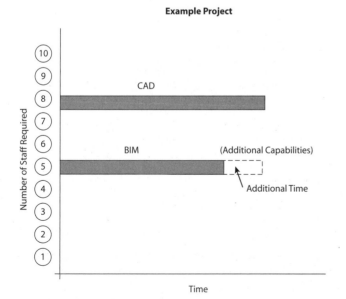

Figure 3.21 BIM efficient team vs. CAD efficient team

The challenge the architecture profession has now is if the contractor is asked to become responsible for even more coordination of projects in creating the BIM model, what then is to become of the architect's role in building design? Although there would still be a need for specification writing, design logic, and client relationships, if the most time-consuming and profitable portion of an architect's fee—the construction documentation phase—is reduced or lost completely, what then are architects creating? The simple truth is that over the past century, the role of architect has been reduced from a studied professional who knew trades, construction methods, and material qualities to a design consultant all in the name of exposure to risk and fear of liability. At the same time general contractors responsible for building the project continue to take on more liability and, as a result, more fee. The general concept is an economic one; those willing to take on more risk will receive more reward. Less risk equals less reward. Let's face it: building structures is not a safe profession. There are risks financially, in working relationships and in reputation, and the need for architectural and construction professionals to collaborate and work with technology and each other toward better virtualization and modeling has never been greater (Figure 3.22). However, the importance of defining these roles is just as important as the collaboration itself.

Figure 3.22 Contractor using BIM in the field

The bigger question is, How does a design model evolve into a fabrication or contractor's model? Obviously, the architect must begin by modeling something and is not necessarily able to produce construction-ready models in the schematic design phase. However, as the number of BIM users in the industry increases and modeling proficiencies skyrocket, the opportunity to add a much higher level of detail to a project earlier is very much a reality. The project team must then decide what information is needed early on and how the team can work together to define it. As a result, there is much more opportunity for architects to create internal efficiencies through working with contractors.

The knowledge of building construction resides in experienced minds. A BIM process uses the contractor as a resource and leverages the tools that BIM is capable of to address constructability issues. Constructability issues and means of constructing it are not the main focus in the schematic design stage; rather they grow in importance with the development of the model and project milestones. This is often referred to as the *level of detail* (LOD) in a model. The LOD in the Model Coordination Plan defines what level of information is needed not only earlier, as discussed, but what can be focused on at a later stage. The AIA produced the AIA E202-2008 document, which aims to describe what level of detail a model should be at during the design and documentation processes. An architect should be open to meeting the team's needs just as a contractor must meet the same expectations. As a user's experience grows in BIM, the following questions will become much more commonplace: Who is modeling the rebar layout for our self-perform scope? Who is modeling the precast embeds and verifying their alignment around the perimeter of a building? Who is modeling the pipe racks, cable trays, and hangers? In the end, it's about creating a model is not only usable for documentation but is also constructible.

Submittals

Submittals, and shop drawings in particular, are a source of continual work for construction managers and architects. It is critical for the contractor to make sure the interpretation of the design intent is accurate with the architect, and for many items in a project, little coordination is required for each. However, the culmination of these items results in quite a large effort. Performance criteria further complicates the issue because engineers create system performance specifications that subcontractors and fabricators then turn into constructible designs. The typical delivery method for submittals have been mailed, faxed, scanned, or posted versions of PDF or CAD information. The medium in which this is accomplished can change in BIM. BIM (especially a BIM process geared toward fabrication) allows shops to potentially utilize the BIM file and generate and send their 3D documentation to the project team for review. The team can then insert the documentation into the model for interference checking, review, and markup (Figure 3.23 and Figure 3.24). This increases the accuracy of the collision detection and keeps long hours at the light table a distant memory. However accurate the model, it is critically important to verify the coordination points prior to linking in shop drawings. Although BIM software in general is not perfect, often file exchanges between different formats will alter some basic settings, so it's always a good rule of thumb before diving in to verify the accuracy by double-checking coordination reference points.

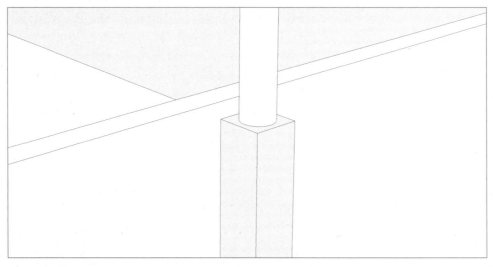

Figure 3.23 Example of early design-level BIM

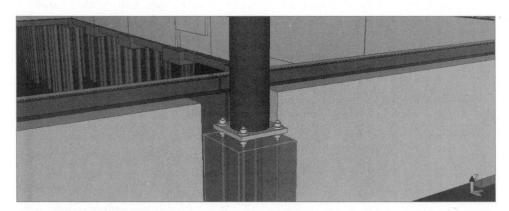

Figure 3.24 Example of fabrication-level BIM

Ideally, when a BIM model is sent to fabrication, the fabricators can use BIM to some level and generate their 3D shop drawings and reissue them to the project team. Project scope such as structural steel, mechanical ductwork, piping, electrical piping, curtain wall, SIPS, precast concrete, and roofing panels are popular. However, other items such as door hardware, finish information (paint, vinyl wall covering, wall base), and other items are typically found in the specification. Although embedding all the specification information into the BIM is possible, it's currently a hybrid process. This hybrid process focuses on using the model to clarify on the more complex systems of building design, which require a greater degree of engineering and coordination. In turn, the model provided by the fabricator can be tested against the composite model for accuracy.

e-SPECS has developed software that is capable of linking model components in a building to a specification. The software allows components to be uniquely linked and specified to components to the MasterFormat specification database. This software represents a process shift, and it's a useful tool for both architects and contractors. Through it, architects can create early skeleton specifications based on the information in their model and then edit, create, and add additional sections as the project progresses. Because this software uses linking, the information is continually shared between Autodesk Revit and e-SPECS and doesn't involve exporting at any time. This software has some limitations, but using this type of functionality within native modeling software will allow users in the future to embed most of the information about a project within the BIM model. This could be extremely valuable not only to fabricators but also to contractors in the field and to facility managers.

Creating a Construction Model

Conceptually, the construction model is an isolated model that allows the contractor to layer information over the architect's and engineer's linked BIM model, without editing their respective models but still being able to access all the necessary information through native modeling software. Although many contractors resolve to create completely separate BIM models or may have a totally different standard of getting the information necessary into the model, I will continue in this example as if we had established in the IE responsibility plan the ability to create a construction model during the design development phase. Creating a linked but separate BIM model is similar to an MEP or structural BIM in that it allows the architect to turn off the model and additional information in all of its views within the project easily, while the information is still available for the architect to review as the design progresses in the composite model. Some of the components typically more important to a contractor than an architect are rebar, site utilities, concrete pours, formwork, equipment accessibility and clearance, lay-down and staging areas, lane closures (with duration), component scheduling (adding detail to existing model parameters), precast panel layout, and steel and large component phasing (dropping in large pieces of equipment and building the shell around it). The following tutorial uses the architect's model and creates a separate but linked construction model to accomplish this.

A Revit construction model is the same as a Revit link that you might have for a mechanical engineer using the same native software. For example, if the architectural team is using Revit Architecture and the MEP contractor is using Revit MEP, then the contractor can link both models into Revit (most often using Revit Architecture software). This allows the contractor to see the objects in a file where information about objects can be added in its most robust form. In addition, updates from the design team can be reloaded into the working composite Revit model. There are two places to house your composite BIM. One is in the native modeling software such as Revit or ArchiCAD. The other is in compilation analysis software such as Navisworks or Solibri. Although you can use the model in Navisworks for linking and testing, using the native modeling software gives you modeling tools that Navisworks lacks. Both tools have unique functions and often are used in tandem throughout a project. To increase efficiencies and limit manually compiling models, linking files saves time and lets the computer manage updating the files. This type of model coordination is required for a few reasons:

- There currently isn't any software available that can do it all in one platform (modeling, analysis, and testing).

- There are too many analysis tools for an all-in-one software tool to do everything associated with managing a BIM model (code checking, CFD, LEED, constructability).

- This hybrid approach between modeling and analysis tools maximizes coordination while not limiting the tools, which can be used.

Linking Models

Linking models prevents the accidental editing, moving, or deleting model elements built by the design team. Linking is a best practice, because it eliminates the ability for the contractor to unknowingly manipulate the model without the approval of the architect, thus preventing costly errors. Using the Coordination Review tool in Revit is another effective way of letting a BIM user know whether linked elements have been shifted during use or because of updates.

Linking in information relevant to a trade lets the model begin to function as a whole. This composite model allows the construction team to find valuable information as well as provide recommended model changes to the design team for elements that are critical to the contractor, such as the exact dimensions of grid lines, levels, wall alignment, and clearances. For example, a linked composite model in native software can be tested for daylight penetration (see Chapter 6), mechanical loads, and many other factors.

The usefulness of creating a construction model with other files linked into a Revit BIM model is that the contractor may input information relevant to his responsibilities, which the design team may then review. What kind of information does the construction manager want to input into the model? In a 350,000 square-foot warehouse, the building will have construction joints for concrete pours; building expansion joints; a significant amount of rebar embedded into the structure to support equipment loads above; and precast panels, which will require steel embeds for welding as well as panels. But just because you can model additional information doesn't necessarily mean you should. Although all these elements can be modeled, it is useful to model only what is applicable or needed. If the concrete is being self-performed by the general contractor or construction manager, then it might be useful to lay out the rebar. If it is being installed by a flatwork subcontractor who is using the services of a structural engineer with their own model or documentation, then it would be best to have the structural engineer model the rebar layout for coordination purposes, especially if it was agreed upon in the model coordination plan. If the model is created by another party, verify that the model can be linked into the composite BIM. This lets the team use the model for reference.

The construction model will be transferred to the field during construction. So, you should discuss the actual setup of input with the construction team using the model and identify which format will be used as the standard for the project. For example, you should ask the following questions prior to building a construction model:

- What format will the in-field personnel be using? Revit models with text notes, Navisworks, or DWF markups for the project?
- What's the expected deliverable to the owner at the end of a project?
- Do the computers on-site have enough horsepower to handle the size of the BIM file?
- What is the field personnel trained on?
- What are the goals of the project?

The answers to these questions will help eliminate confusion and streamline communication while creating a construction model, prior to using it in the field.

You can set up BIM for use in the field in a number of ways, and ideally they all begin with creating a construction model. Some prefer to link in DWF comments on 2D drawings into the RVT file and use this as a means of issuing RFIs through-out construction. Highly integrated teams use the same model to copy in comments and coordinate construction on site. Others work in the same file through colocation, which requires saving to a central file over the Internet (which can be time-consuming, depending on file size and data transfer speed). Some construction managers use Navisworks to manage all correspondence throughout a project. Keep in mind that there is no right or wrong way to manage a BIM project in the field. Conceptually, the idea is to create the information in a format that will allow you to use, extract or link it later into a Record BIM.

In this book, you will examine two ways of managing a BIM model in the field. In the first, you will use 2D and 3D DWF files, which can be used to import all the comments and markups associated with the drawings into the native Revit file. In the second way, you will use a Navisworks file, which I will discuss in Chapter 5. To begin, I will show how to set up a separate construction model file.

Creating a Separate Construction Model

1. Create a new Revit file, and save it as Construction.rvt.

2. Choose File > Import/Link > Revit (Figure 3.25).

3. Locate the architect's model, Example-50% dd.rvt.

4. Save the file as Construction.rvt.

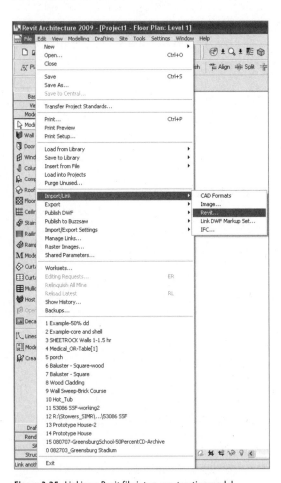

Figure 3.25 Linking a Revit file into a construction model

 Note: You have to close out of the model you are trying to link in to your construction file. Otherwise, you'll receive an error message and cannot successfully load the file.

When you have imported the model, you will be unable to move any of the components in the linked file. This is a feature that keeps you from inadvertently changing the architect's and engineer's design. When working on files submitted by architects and

engineers, linking is a best practice, because it removes the chance for human error, as discussed earlier. In addition, if the building is accidentally moved, it will shift the entire building, not just the components, and should appear visibly out of alignment.

To select model components, use the Tab key to cycle through selection sets. The default selection is the entire model, but by cycling through using the Tab key, you can select elements and then check their properties by clicking the Element Properties button in the top left of the Revit window (Figure 3.26).

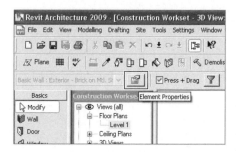

Figure 3.26 The Element Properties tool

Now you can begin layering constructability information on top of the Revit model by creating a new workset.

Adding Constructability Information

1. In the new Construction.rvt file, select File > Workset.
2. Click New, and name the new workset **Construction**. Select Visible by Default in All Views. Click OK (Figure 3.27).

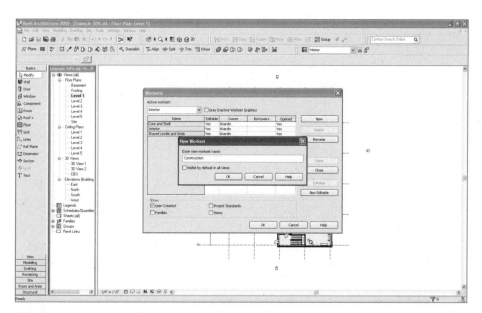

Figure 3.27 Creating a new workset

3. If it is not the default, click Yes to make Construction the active workset.

4. Open the Construction workset, and open the Level 1 plan view.

    In the new Construction workset, you will create two new floors on top of the generic floor. This is where you will recommend to the design team how the concrete pour will be sequenced.

Note: To change the visibility setting within Revit to a more detailed level, simply type **VP** or select View > View Properties. Under the detail level, change the view from Course to Fine. This will enable you to see the walls layers prior to sketching the floor outline.

5. Click the Floor tool on the Basics tab (Figure 3.28).

6. Click the Lines tool (Figure 3.29), and begin tracing the floor, as shown in Figure 3.30.

Note: The other useful tool in the menu is the Pick Wall option, which allows the user to pick the walls associated with the floor's boundaries.

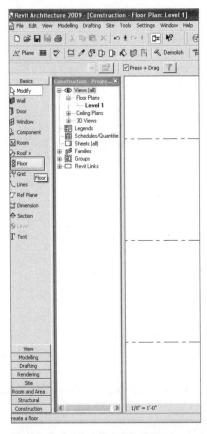

Figure 3.28 Creating a floor in Revit

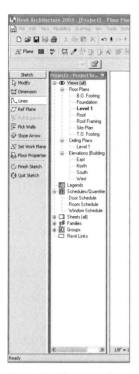

Figure 3.29 The Lines tool

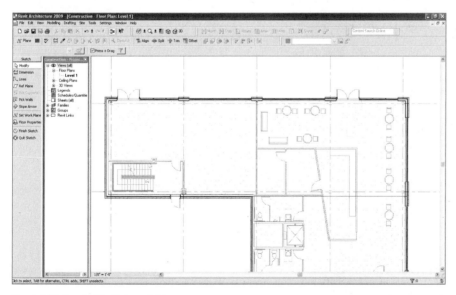

Figure 3.30 Drawing the first pour joint

7. Click the icon Floor Properties (Figure 3.31), and change the Type drop-down list to 4″ Concrete Slab (Figure 3.32).

8. When the sketch is completed, click Finish Sketch.

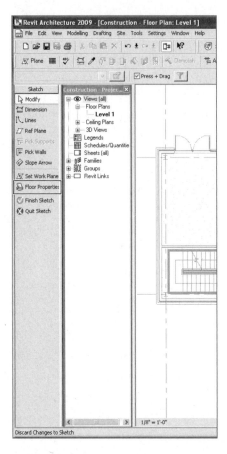

Figure 3.31 Floor Properties icon

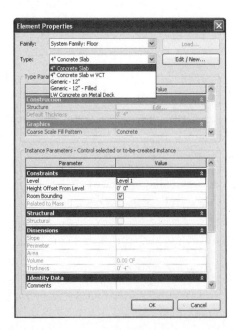

Figure 3.32 Changing floor properties to a 4-inch slab.

Continue the tutorial by repeating the previous steps to create the adjacent floor slab pour.

Although this book cannot possibly get into every detail of how to create a construction BIM model, this floor layout would evolve and eventually contain embed dowels at this joint, rebar layouts, and edge-of-slab profiles. It could be phased depending on the type of software used to generate the 4D scheduling. For the example in this book, you will not add this information. However, know that this information could be very useful in the field and should be input into the construction model for field reference later if needed.

As linking models delineates one model from other models, you can generate this linked and unique construction model in house. It is best used to coordinate internal efforts while collaborating with the architect's and engineer's design model to avoid altering the model. In a typical project, this model would be sent to the AE team to review, and then comments would be generated from the structural engineer and architectural teams. Eventually this model becomes the construction BIM model as the project nears the construction phase of the project.

Using the separate construction model is a great in-house coordination tool. The following are the advantages to creating one:

- Being able to use the modeling interface to verify the accuracy and coordination of elements, while still playing in the same sandbox as the rest of the team
- Being able to reload updated links, allowing the construction manager to see design changes quickly
- Being able to create text and callouts that are useful to the construction manager and the design team
- Creates views and text that remain specific to the construction manager
- The ability to pass it off to other project management members, who might not have been intimately involved in the design of the project, to see coordination issues and to better understand changes

Design Review Tool

Revit was designed to function as modeling or model-producing software, which often results in large file sizes. Therefore, its ability to function as a usable field tool is somewhat limited. A best practice is to have the construction model on hand at a job site, but certain pieces of software are good at certain tasks, and some aren't. So, as the project progresses to the field, you can use other tools to coordinate the construction of a project without having to force Revit or other modeling software to work in the field. One of these tools is Autodesk Design Review.

Originally this tool was merely a viewer with limited capabilities, both model-related and for commenting. With the latest release of Design Review, however, certain functionality makes it much more useful than it was in the past. In the following tutorial, you will learn how to take both sheet drawing information and the model and combine them into two files that allow commenting, tracking, and linking to take place. At the date of this publication, the ability to create a document in Design Reviewer that contains both the 3D model and the 2D sheet files doesn't exist, so for this example you will create two files. One will be the 3D file that allows for commenting but can't be imported back into the Revit model. The other is a 2D DWF file, which allows comments and markups to be imported back into the Revit file.

Design Review (www.autodesk.com/designreview) either can be installed in tandem with the Autodesk Revit suite or can be downloaded from the Autodesk website.

Exporting the Revit Model to Design Review

1. Launch Revit, and open the file Example-50% dd.rvt.

2. In the view browser, open the sheets directory. You should see a list of sheets that have been created since the earlier model.

3. Open the sheet labeled A100 - Level 1 Floor Plan.

4. Select File > Publish DWF > 2D DWF (Figure 3.33). This opens the Publish DWF 2D dialog box (Figure 3.34).

5. Select DWF Files (*.dwf) as the type of file to export, and change the view range from Current View to Selected Views/Sheets.

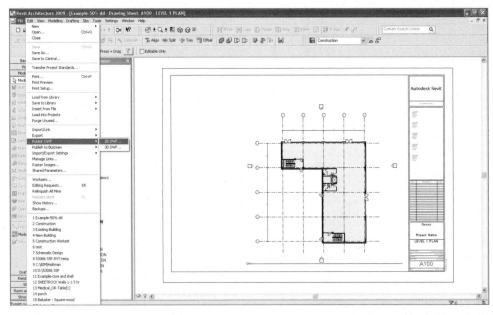

Figure 3.33 Exporting sheet drawings to Design Review

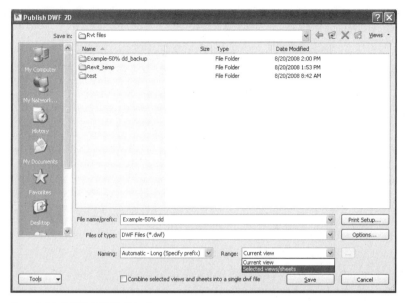

Figure 3.34 Publish DWF 2D dialog box

Note: The difference between DWF files and DWFX files is that the DWFX files are supported by Microsoft XPS Viewer that comes preloaded with Windows Vista. These can be changed, but some programs won't open the DWFX file format; for this reason, it's better to stick to the DWF file type unless you need the performance and are running Windows Vista.

6. Select the Combine Selected Views and Sheets into a Single DWF File box.

7. Click the Browse button ([...]) to the right of the Range drop-down list. This opens the View/Sheet Set dialog box.

8. Specify all the sheets in this project. If you want to see a 3D view or other view that isn't on the sheets, you can specify which views you want to be part of the DWF here.

9. Click OK, and click No to avoid saving the sheet selection set.

10. Specify where you want the DWF file to be saved, and click Save.

When dealing with RFIs and in-field document management, it's important to have as many resources available to you as possible, especially ones that meet your needs. This tutorial shows a limitation in the inability to be able to mark up the 3D DWF and reinsert those comments into the Revit model. A better option might be tools that Navisworks provides, which will be outlined in Chapter 5.

Open the newly created DWF file to begin to familiarize yourself with the Design Review interface (Figure 3.35).

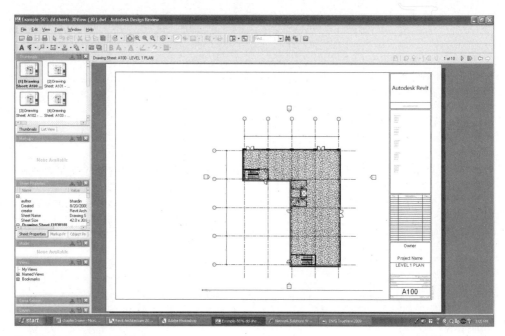

Figure 3.35 Autodesk Design Review

I will now show how to comment on a DWF 2D sheet set and import those comments into the Revit model.

To begin, open the sheet A100. Click the Callout drop-down list, and select the Rectangle Callout, Revision Polycloud (Figure 3.36).

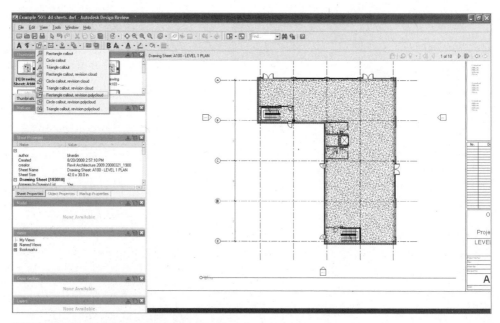

Figure 3.36 Selecting the revision callout type

Commenting on a 2D DWF Sheet Set

1. Trace a box around the south stair tower.

2. Click near the starting point to close the callout.

3. Move the mouse to locate where you desire the text portion of the callout to be located.

4. In the text field, create a callout that asks, "What is the rating for this stair enclosure?"(Figure 3.37).

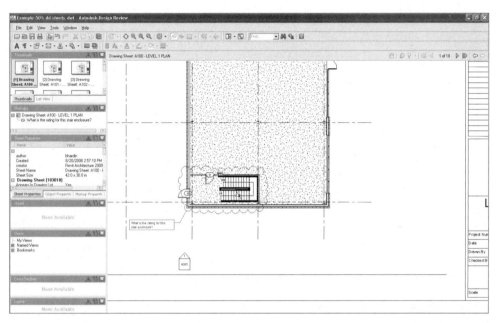

Figure 3.37 Creating a clarification in Design Review

You should now be able to see the new markup under the Markup heading to the left of the screen. The details of when the callout was created, as well as the option to assign a status and notes to the callout, are available when the markup is clicked and the Markup Properties dialog box appears. This, of course, shows the basic functions of the DWF reviewer, and the BIM user may choose to add a preceding RFI number or DD comment number prior to the text field as a reference.

You will now import the file into Autodesk Revit. To begin, save the DWF Design Review file, and open the Example-50% dd.rvt file.

Importing a File with Comments into Revit

1. When in Revit, select File > Import/Link > Link DWF Markup Set (Figure 3.38).

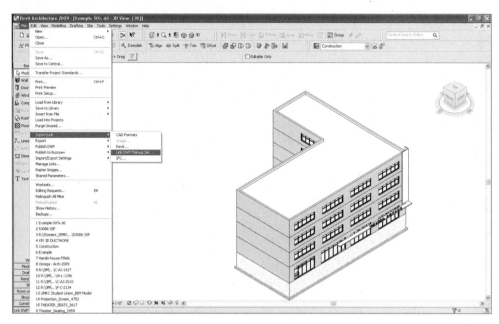

Figure 3.38 Importing DWF Design Review comments

2. Specify the location of the DWF file, and click OK to link the sheet to the model.

3. Now open sheet A100 in the Revit model, and you should see the comments created from the DWF design reviewer tool (Figure 3.39).

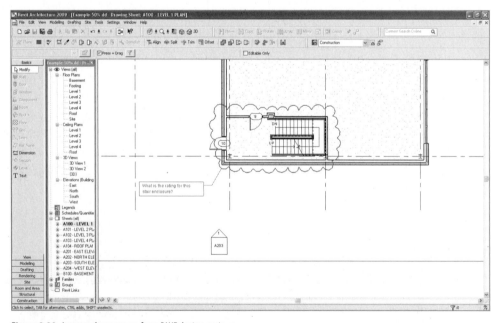

Figure 3.39 Imported comments from DWF design reviewer

Utilizing this feature is an effective way of managing coordination issues during design and during construction, with the exception of being able to comment on the 3D view itself. A tool that allows both 2D sheet documentation and Revit or DWF files to be used in the same file is Adobe Acrobat Professional Extended. Adobe is very helpful and offers a powerful array of tools, but it does not link comments into Revit. In addition, Adobe offers a tool called Review Tracker, which will notify users of an update to specified Adobe files similar to an RSS feed. Adobe could easily become the preferred method of managing comments and tracking them; however, the limitations between Revit and Adobe currently prohibit complete coordination.

Trade Coordination

A major portion of the construction manager's responsibility is coordinating multiple trades. Trade coordination involves working and communicating with subcontractors, supervisors, material suppliers, fabricators, and specialty equipment suppliers, among others. In addition to juggling the scheduling, managing the budget, sorting through constructability issues, and managing relationships, the construction manager is also responsible for coordinating who is doing what work on a project. This is a daunting task, especially when the scale and complexity of the project escalates.

In the past, trade coordination was accomplished either on light tables with plans overlaid on each other or through 2D CAD drawings that were referenced on top of each other in the computer during the initial design stages. These CAD drawings, which lack a z-axis, allow for mistakes in interpretation. For them to be completely accurate, the top and bottom elevations of all the components in a project need to be shown in order to coordinate these layered CAD files.

Although many projects establish a plenum space, below-floor elevation space, under-slab elevations, and chase spaces in which to run equipment, there is often no real way to determine the actual dimensions of equipment as it reduces in size from one floor to another, the layout of the supporting raised floor columns, or the rebar layout in a floor that is to be core drilled. The only real way to accomplish a useful coordination model is to create a composite model in which all files are 3D, linked, and intelligent during the design phase (Figure 3.40). Trade coordination is one of the areas where BIM really shines.

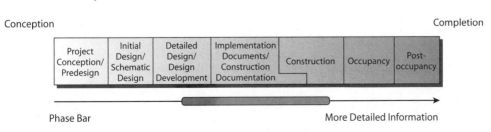

Conception   Completion

| Project Conception/ Predesign | Initial Design/ Schematic Design | Detailed Design/ Design Development | Implementation Documents/ Construction Documentation | Construction | Occupancy | Post-occupancy |

Phase Bar   More Detailed Information

Figure 3.40 Trade coordination begins in design and continues though construction.

Clash Detection and Reporting

One of the major factors leading to the building information modeling movement for contractors was the drive to develop clash detection functionality between models. The degree of accuracy and the ability to layer multiple data sets and models into one file are new in the construction industry. These capabilities show where BIM can provide a tool that 2D data can't touch.

At the onset of a project, you can run entire models against other models to see what the scope of interference is. Virtually anything in the model can be tested against another set of objects, elements, or selection criteria. As the number of reported clashes diminishes, the areas that are being tested can be narrowed down; they can be avoided if there are known issues that are to be resolved later in the project timeline.

Not only do the clash reports need to be generated and distributed to the project team, but these conflicts also need to be resolved!

AISC Guiding the Way

By Erika Winters Downey

The American Institute of Steel Construction (AISC) has been a driving force behind many of the BIM efforts in the industry today. It has been promoting the adoption of BIM in steel construction for more than a decade. Erika Winters Downey, S.E. AISC Great Plains regional engineer, has provided this sidebar to elaborate on AISC's role.

The structural steel industry experienced a significant improvement in productivity during the past two decades of the 20th century as a result of improvements in the mill process of producing structural steel. The average number of hours required at the manufacturing mill per ton of structural steel plummeted from 12 to 0.5 during this period. Following these improvements, the industry turned its attention to other aspects of the supply chain for fabricated structural steel in hopes of identifying specific activities that could yield similar productivity improvements.

In 1998, AISC analyzed the traditional workflow processes in steel-framed building projects and identified inefficiencies when 2D drawings are manually transferred between parties instead of electronically transferring the data. AISC evaluated several existing file formats already commercially available and chose CIS/2 as the best format for neutral file transfer. CIS/2, or CIMSteel Integration Standards/Version 2, is a U.K.-based data dictionary and file format that AISC chose because of its robustness in terms of its abilities to assign intelligent entities within a structural model and then manage and track changes to it. In 2000, major software firms in the United States agreed to incorporate CIS/2 technology into their programs in exchange for a three-year moratorium on changes to the CIS/2 standard.

AISC Guiding the Way *(continued)*

As a result, project teams can save time by electronically transferring data from a structural BIM to a manufacturing model, rather than starting a new model from scratch using 2D design drawings. Programs at the fabrication level allow for electronic development and review of shop drawings. The fact that structural steel is fabricated in centralized locations, remote from job sites, inherently allows it to adopt automation in the workflow. At the fabrication shop, information from the 3D manufacturing model is sent to computer numerically controlled (CNC) machines on their beam lines. This process integrates well into a lean construction model utilizing 3D design of MEP systems, cladding, and architecture.

In the 2005 AISC *Manual of Steel Construction*, the Code of Standard Practice debuted Appendix A, "Digital Building Product Models." When specified, the design model will govern over architectural and structural design drawings. The manufacturing model will govern over shop and erection drawings. It also sets procedures for the logical product model, which encompasses the analysis model, the design model, and the manufacturing model.

AISC continues to support CIS/2 by ensuring its integration into the larger AEC industry. The first step in doing so was to develop a translator that would allow a CIS/2-to-IFC exchange. This allows steel-framed structures to be integrated into the standard that is internationally recognized. With a CIS/2-to-IFC translator, CIS/2 became capable of exporting structural steel models to IFC-compliant building software. Additionally, AISC began to participate in the National BIM Standard (NBIMS), which was an undertaking of the buildingSMART Alliance and the National Institute of Building Science (NIBS). AISC's work with NBIMS has revolved primarily around defining how and when information is exchanged in the structural steel design process.

Navisworks Clash Detective

The following tutorial uses a new mechanical ductwork model and runs clash detection against the structural steel. It shows how to generate a clash detection report, distribute it to a project team using HTML or a Navisworks viewer file, and develop a resolution plan that shows the clashes and tracks responsibility for resolving them.

Using Navisworks Clash Detective

1. Launch Navisworks, and select File > Open > Example-50% dd.nwd.
2. Click the Clash Detective button (Figure 3.41), and dock the Clash Detective palette by clicking the pushpin. The Clash Detective tool tests 3D (or CAD with *z* coordinates) information in the left pane against information in the right pane.

Figure 3.41 Clash Detective button

3. Append the mechanical model to the architectural model in Navisworks.

Note: Revit Architecture opens Revit mechanical models, but the ability to model using the mechanical interface is unique to Revit MEP. The same is true for Revit structure files as well. In other words, a model created in Revit Architecture must continue to be modeled in Revit Architecture. A model built in Revit MEP must continue to be created in Revit MEP.

4. Select File > Append, and choose the file `FG-HVAC-04.nwd` file (Figure 3.42).

Figure 3.42 Append command in Navisworks

Note: Although this tutorial tests two models against each other, all elements can be selected and tested against each other. For example, a search set of the structural steel can be created and then tested against the concrete floors of the same model. As an additional exercise, create search sets of both the structural steel and the concrete floor from the architectural model, and run an additional clash detection batch. This will help you familiarize yourself with the capabilities of the clash detection function in Navisworks.

You now see that under the clash detective reporting windows there are two model files (Figure 3.43).

5. Select the FG-HVAC-04.nwd file in the left window, and select the Example-50% dd.nwd file in the right window (Figure 3.44).

6. Make sure that the Self Intersect check box is cleared on both sides.

7. Set Type to Hard at the bottom of the Clash Detective palette.

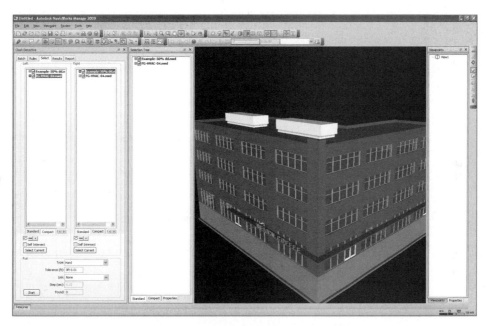

Figure 3.43 Separate files are identified in the Navisworks clash detective windows

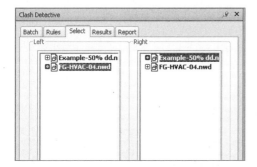

Figure 3.44 Comparing two models against each other

Note: A *hard clash* is the physical intersection of two 3D components, whereas a *clearance clash* will report if components are within a specified dimension of each other. Duplicates identify two objects that are identical in type and position.

8. Set Tolerance to **0ft 0.01**. This is the degree of interference that is acceptable. It defines a rule for generating the clash report.

9. Click the Start button to generate the report.

10. Click Save.

If this example has been completed correctly, you should have 751 clashes, which you can see by clicking the Select tab and looking at the bottom of the screen (Figure 3.45).

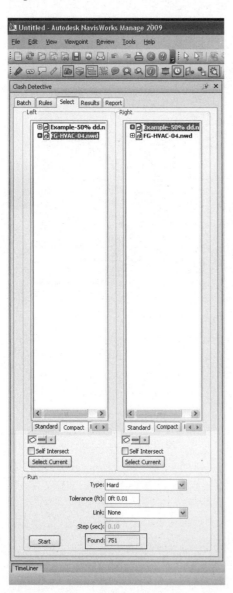

Figure 3.45 Found clashes using Navisworks

By clicking each clash on the Results tab, Navis will automatically zoom into the area of the clash in the 3D viewer. You can use the Display tool on the right side of the viewer to change how clashes are displayed. Click through the clashes, and see how the Auto Zoom, Hide Other, and Dim Other display features can be toggled. This will help you clarify exactly what you are seeing. Keep in mind that you can use the Orbit, Pan, and Zoom tools at any time when viewing a clash report in the clash report view.

Other elements on the Results tab are a series of headings titled Name, Status, Distance, Description, Found, Approved, Approved By, Clash Point, Start, End, and Event. These are described in Chapter 4, which discusses using the clash detection report to generate actions and to update the report.

Save this file as `Example-50% clash.nwd` when you have completed the tutorial, because you will use this NWD file in the next chapter to begin resolving and updating the clashes.

Case Study: Managing Construction of a Signature Bridge with BIM

Standing tall as a vital transportation link in the Bay Area, the current construction work on the San Francisco–Oakland Bay Bridge (SFOBB) is a premier example of renewal and investment in America's critical bridge infrastructure. It is also an excellent example of BIM applications making a difference in a large construction program. In this case study we outline the processes and tools used across the delivery team to improve quality, reduce risk, and enhance communications with the traveling public.

Enabling nearly 300,000 vehicles crossing daily, the two spans of the SFOBB are among the busiest bridges in the nation. In the aftermath of the devastating 1989 Loma Prieta earthquake, the California Department of Transportation (Caltrans) determined that the East Span would be replaced with a new structure and the West Span would be seismically retrofitted. Both projects would be completed with the structures remaining open to traffic. This project is among the largest public works projects in California history, costing $6.3 billion, and features the world's largest self-anchored suspension span.

Based on the early success of 3D model renderings of the new bridge during the extensive environmental approval phase, Caltrans requested that Parsons Brinckerhoff (PB) develop an updated 3D model of the entire span reflecting the completed design for use in its new outreach strategy in led by the Bay Area Toll Authority (BATA). At the time, many of the design activities were being completed in standard 2D CAD applications. PB produced an accurate 3D digital model of the entire East Span corridor, including the existing bridge, temporary structures, and the future new span renderings using Autodesk 3ds MAX software (see Figure 2.36).

IMAGE COURTESY OF PARSONS BRINCKERHOFF.

Figure 3.46 Rendering of new bridge work and temporary detour route.

To extend the utility of the initial modeling activity, PB went on to create a 4D project model using Autodesk Navisworks. Components and groups of components within the 3D model were linked to corresponding construction activities in a master Primavera-based contract schedule. The resulting interactive 3D simulation of the entire project shows construction activities including staging and equipment moves over time in an effective and realistic way (see Figure 3.47). For the BATA project team, this was the first implementation of a 4D process during construction, and the ease of use enabled them to immediately leverage the benefits during project review meetings.

The BIM project model achieves several objectives:

- Better inform the design and construction teams about planned construction processes

- Help communicate to decision makers when certain activities will take place and the relationship among key milestones

- Foster more collaboration among project partners and stakeholders

- Enable the public to clearly see what the project will look like over the course of construction

- Facilitate media campaigns that effectively communicate planned closures and traffic detours

Figure 3.47 Stills from completed sequencing animation.

Early acceptance led to an iterative model revision process within an overall model management framework. The project model was maintained as designs were refined and the schedules updated. The model was extended to include fabrication and delivery of deck and tower sections for the self-anchored suspension (SAS) portion of the bridge, critical activities for overall construction scheduling (see Figure 3.48). The detailing of components and complexity of geometry has expanded with use. Currently the model links to over 3,000 construction activities in the multiple contracts, and the model itself has grown to over 800 million polygons.

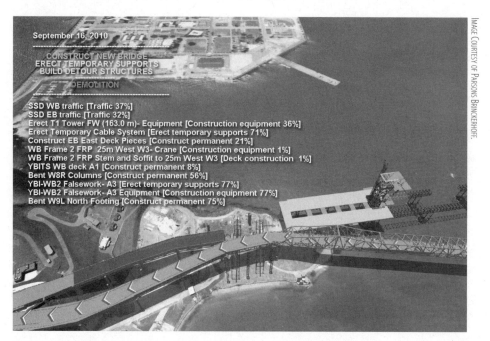

Figure 3.48 ■ BIM of the bridge showing daily construction activity tied to rendered components.

The composite model has become an integral part of several notable achievements. In one example the clear depiction of what were initially overlapping activities in the same work area among multiple contractors' construction schedules helped resolve conflicts in confined areas with limited access on Yerba Buena Island between the two spans. Within the model simulation, multiple planned activities can be shown together for any day during the life of the project. This approach to construction sequencing is being embraced by clients and team members at every level of the project including the Caltrans scheduler, Toll Bridge Program Oversight Committee (TBPOC), and even the executive committee responsible for making major decisions on the project.

The project work has created many new opportunities for Caltrans and BATA to explain the project to non-technical stakeholders, and has made stakeholder communication easier and more comprehensive. Caltrans and BATA have committed to the continued use of visualization and 4D modeling to illustrate construction sequencing for several more key components of the new bridge, including a detailed visualization of the construction of the SAS portion of the bridge.

The structural and MEP components for the bridge were modeled in 3D and compared for potential design conflicts including interface issues between the two construction document packages being developed in parallel.

The model-based methods and associated used on this project has led to discussions by Caltrans leadership on the inclusion of 4D modeling and digital prototyping as standard components of their project development process for their large infrastructure projects.

The following specific construction challenges were managed more effectively through innovative use of BIM processes.

Challenge: There are three prime contracts with work areas on Yerba Buena Island, the small island (only 300 acres) that serves as the connection between the east and west spans of the SFOBB. The team needed a way to illustrate the work areas for each contract, as well as communicate the limited geographic area in which to stockpile materials and conduct construction operations. Contractors needed to be informed about the state of construction adjacent to their work areas at the beginning of the contract, as well as the location of environmentally sensitive areas (ESA).

Solution: Using a 2D sheet file of the contract work area boundaries, 3D splines were created and formed to fit the base digital terrain model (DTM) in 3D. A high resolution aerial image was draped over the terrain model, and boundaries were color coded to illustrate each contractor work area. The work area geometry was imported into the 4D model, and images were submitted to the client for different points in time during the project construction.

Challenge: The 150-meter tall suspension tower for the Bay Bridge has extensive internal mechanical, electrical, and plumbing (MEP) design elements, in addition to a seismically innovative structural design. Because of the size and complexity of the tower, the project team needed a way to perform design validation for the variety of engineering disciplines. PB designed the MEP elements of the tower, while a Joint Venture team, TYLin/Moffat Nichol, designed the structural systems. Although the structural design was created in 3D using Dassault SolidWorks, proprietary considerations in project contracts made it an issue to share design information between team members.

Solution: The design team was able to provide their 3D structural design in an Autodesk Navisworks .NWD file format without sharing proprietary design information.

Navisworks then was able to import both the .NWD file exported from SolidWorks with the MEP design .NWD file exported from Autodesk 3ds Max to perform clash detection operations on both design models. Initial clash processes yielded multiple issues, and electrical designers were provided with reports generated from Navisworks that clearly demonstrated the areas of conflict.

Challenge: The 2,047 foot self anchored suspension span of the new SFOBB will be the largest of its kind in the world. At a cost of $1.45 billion, the span is being built under the largest infrastructure contract ever awarded in California. The steel deck and tower components will be prefabricated in China and shipped to the project site, creating monumental logistical and managerial challenges for the project team. An effective method to visualize and understand the interaction between the fabrication and construction operations was needed.

Solution: Detailed structural models for the bridge components were created using original 2D design data. PB then used construction schedules for the project in coordination with a procedural fabrication schedule to create a 4D model for the prefabricated deck and tower components that simulated segment and lift assembly, shipment and inspection status, as well as placement of the prefabricated components on falseworks at the project site (Figure 3.49).

IMAGE COURTESY OF PARSONS BRINCKERHOFF.

Figure 3.49 Rendering showing phased construction activities and break down of the old structure being phased into the new structure.

Challenge: Both directions of the SFOBB were closed for Labor Day Weekend 2007 to allow workers to replace a 350-foot section of roadway. As California's busiest bridge an enormous public outreach effort was needed to change area motorist's behavior during the closure.

Solution: 600,000 fliers were printed, hundreds of Electronic Message Signs were placed, radio ads and television commercials ran, presentations aired before the coming attractions in movie theaters throughout California, and giant banners were placed in the Bay Area's airports. A major element in this endeavor was a set of 3D model renderings and animations generated from the project model and used to quickly explain step by step exactly what was going to happen. At the time of the closure, surrounding traffic was so minimal the bridge was closed almost an hour before scheduled.

Challenge: There is an enormous interest in what the new bay bridge will look like in its future environment. Numerous images and animations of the new bridge have been released to the public through local media. These resources are valuable, and can answer many questions, but seeing the bridge in its context from any vantage point in an interactive real time 3D model is the ultimate way to quickly show and explain this future icon to the general public.

Solution: A less detailed model of the New Bay Bridge was generated and imported into Google Earth. Inside Google Earth, a model of the completed bridge exists in its geographic location. Along with the ability to fly around and view the model from anywhere, the model has placemark links to information on the new bridge and upcoming closures of the existing bridge.

Conclusion

This chapter evaluated the tools available for building information modeling during the construction phase of a project. The next chapter analyzes how to update these tools to make them effective throughout a project. Current BIM tools are useful and early adopters will assuredly reap the rewards of future releases and updates. The advantages of working BIM into a construction management process now will pay dividends later, as software and technologies continue to develop, making the construction manager increasingly effective. Software will continue to develop at a rapid pace over the next decade, and users who have implemented BIM into their process will be at a significant advantage over those who have not. Holding off now and attempting to adopt many new BIM tool offerings later will necessitate a lot of catching up.

Utilizing BIM during construction is when many of its values shine. You may have noticed that for the BIM process to work, there are different software applications for different purposes rather than a streamlined and connected one; however, keep in mind that a BIM model—as opposed to CAD files—contains large amounts of usable data and that each piece of software analyzes and sorts the data in a different way. As you go through this book, you will learn how and when these varying tools should be used and what software is handing off information to the next tool.

In this chapter, you began with an architect's model in Revit, linked it to Navisworks for scheduling, created a linked Revit model in which to layer constructability information, handed the Revit information off to a DWF file to begin setting up the in-field RFI process, and then returned to the Navisworks file to run clash detection. In essence, there are many tools required to do the job in virtual construction. Using and maintaining the BIM model will continue to be a learning process, as there is an existing array of tools, and ideally there will be more coordination among them moving forward.

BIM and Updates

This chapter further explores how to utilize BIM in construction, how to use BIM tools to provide updates, and how to use the BIM file to track design changes. Continuing the tutorial examples started in the earlier chapters, this chapter explores how to keep the model current by updating information based on the systems covered thus far. Specifically, this chapter covers the following topics:

Overview of BIM and Updates

In this book, you've seen how BIM can be a useful tool in the preconstruction and early construction phases. But how does it stay a valuable tool throughout a project? One of the answers to this question is through *updating*. Toward the end of preconstruction and at the beginning of the construction phase of a project, the flow of information in a project increases. Specifically, addenda, supplemental drawing information, and submittals, to name a few types of information, start becoming available more rapidly than before. The success of the project starting correctly is usually a direct result of how well this rapid influx of information is managed, tracked, and distributed. Many times this involves the use of a *gatekeeper*, a single person or team that is responsible for managing the information coming in and then distributing it to the rest of the team. The role of gatekeeper may be a project manager, support staff, model manager, or other personnel. The information distribution's success depends on the ability of the gatekeeper to communicate to all parties, to make sure that the correct data is distributed to the relevant people, and to manage the documentation. Managing the documentation includes tracking where it came from, the date it was created, the issue or question involved, and who is responsible for its resolution. Many times it is this additional information that changes bid dates, prices, scope, and responsibilities and that can add a layer of complication and difficulty to a project.

BIM is a single source for information in a project; with it, you can get information and view the latest design changes relatively easily. Although the tools associated with building information modeling have to date been mostly focused on the architecture and engineering sides of things, momentum is picking up for contractors. For this reason, many of the ways BIM is used during the preconstruction and construction phases of a project have improved much more than once believed possible. Modern BIM solutions continue to involve a mix of software systems to accomplish varying tasks. To be effective, these tools must not only accomplish the tasks they were intended for, but the files must be set up so that they can be updated efficiently (such as model linking, as opposed to exporting). The pressure during preconstruction typically intensifies the closer a project gets to the construction phase. This means that it's critical for the tool sets used to focus on accomplishing the necessary tasks and to limit repetitive work. The term *interoperability* applies directly to this ability or inability to transfer information during the life cycle of a building from designing it to decommissioning to future projects (Figure 4.1). Menial and repetitive tasks contribute directly to wasted time on a project and are not nearly as effective as the tools that allow for information to be linked, pathed, analyzed, and updated quickly.

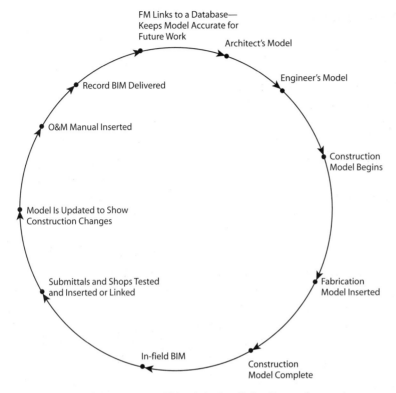

FM Links to a Database—Keeps Model Accurate for Future Work

Architect's Model

Engineer's Model

Record BIM Delivered

O&M Manual Inserted

Construction Model Begins

Model Is Updated to Show Construction Changes

Submittals and Shops Tested and Inserted or Linked

Fabrication Model Inserted

In-field BIM

Construction Model Complete

Figure 4.1 BIM evolution in a project. BIM evolution is cyclical and impacts future project additions and renovations.

This book presents the case for a BIM evolution to a composite model. BIM during the updating and development (design development to construction documentation) stages may run into the following issues:

Challenge: Existing project timelines have limited time to accomplish necessary BIM tasks.

Solution: Log the amount of time needed for future scheduling and evaluate which tools are most needed and which aren't as critical.

Challenge: Low experience levels mean additional time and projects until users become efficient and familiar with BIM both as a technology and as a new process.

Solution: Create an in-house FTP or guide for users to share experiences and improve faster. Additionally, verify in-house resources are adequate for new personnel.

Challenge: Old processes are used as the benchmark instead of new processes resulting in better project outcome.

Solution: Using old processes based on old technology doesn't make sense for new tools and technology. Use these new processes as a relative benchmark, not old processes.

Challenge: Creating and compiling the model, testing it, and addressing issues all take more time than relying largely on field personnel to address constructability issues in the field.

Solution: BIM leverages computers as an additional analysis resource. Addressing issues in the field is often a more costly strategy. Use BIM technology as an extra player in combination with other analysis strategies.

BIM is linear in its process of refinement and development and cyclical in how it is tested, revised, and resubmitted. Take, for instance, how a structural BIM might evolve:

1. An initial design-level BIM model created by the architect or engineer is tested in design and layout and eventually becomes...

2. An analysis and testing model where some detailing has been added. After further testing and refinement, it becomes or replaced by...

3. A fabrication model, which is fully detailed and dimensionally accurate. This is rigorously analyzed and is then tested against or becomes...

4. The shop drawings. Later, if there are changes made in the field, the changes are made to the detailed model reloaded into the project and...

5. The record BIM is delivered to the facility manager who updates any information associated with the structure for future projects.

Layered constructability information increases in sophistication and detail as the project nears the construction phase (Figure 4.2). In essence, BIM represents an evolution in and of itself as a design and construction tool. However, it is during this period of heavy coordination in which a BIM project team will be tested. This is especially critical in regard to updating. Updating estimates, clash detection reports, schedule animations, and constructability models all directly correlate to the quality and output of a project. If the information is outdated, then the tool becomes useless. To some extent, the ship is steering itself, as they say. The idea with BIM is to coordinate and keep coordinating project information throughout the project's life cycle, especially during updates.

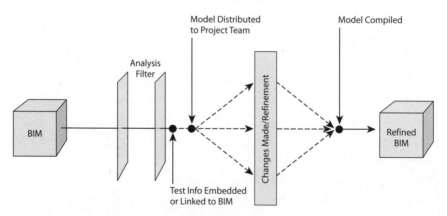

Figure 4.2 Diagram of analysis and refinement to a BIM process

BIM and Prebid

Typically the data that is transferred among the project teams during the prebid phase of a project is a combination of PDFs, scan files (TIF, JPG, and BMP), CAD drawings, and paper documentation. As a project nears bid day, the estimator is responsible for making sure that the information that they are quantifying (PDFs or CAD drawings) is the latest set of documentation. They then rely heavily on the specifications to clarify the scope of work, degree of finish, material type, and quantity to verify with the sub-contractors what is included in their price as they bid or are working toward issuing a GMP or budget. It is not uncommon for estimators to have to make assumptions on a scope of work because of incomplete construction documents or specifications. Based on my experience, often you'll have either overlapping scopes or all of the scopes are not covered when the project budget is issued, because information is missed or added without notification and therefore the details remain unclear. Although this is a relatively common dilemma in the construction industry, it is usually resolved prior to actual construction and addressed in the construction contingency of a project.

The difficulty in hard bid projects (to general contractors in particular) is the disadvantage presented by scope or coordination issues that arise in the field that either were missed or were not coordinated in the contract documents. Typically a general contractor will gain an understanding and level of completeness of a project and, basing a number on that documentation, will carry a large or small amount for contingency depending on the project scale and complexity. Contingencies have been an industry standard for addressing issues that arise in the field for some time, and there have been two schools of thought as to their validity. The first is that contingencies are additional money provided by the owner to essentially pay for mistakes in the documentation and coordination of a project, and the second defines contingencies as essentially quantifying the unknown based on experience and precedent. The Association for the Advancement of Cost Engineering (AACE) defines *contingency* as follows:

> *An amount added to an estimate to allow for items, conditions, or events for which the state, occurrence, or effect is uncertain and that experience shows will likely result, in aggregate, in additional costs. Typically estimated using statistical analysis or judgment based on past asset or project experience. Contingency usually excludes:*

- *Major scope changes such as changes in end product specification, capacities, building sizes, and location of the asset or project;*

- *Extraordinary events such as major strikes and natural disasters;*

- *Management reserves; and*

- *Escalation and currency effects.*

Some of the items, conditions, or events for which the state, occurrence, and/or effect is uncertain include, but are not limited to, planning and estimating errors and omissions, minor price fluctuations other than general escalation, design developments and changes within the scope, and variations in market and environmental conditions. Contingency is generally included in most estimates and is expected to be expended.

"Cost Engineering Terminology," Recommended
Practice 10S-90, AACE International, WV, rev. 2007

Although it is nearly impossible to perfectly coordinate construction documents with or without BIM, a BIM specific process can help more clearly define project scope and budget issues to subcontractors and therefore potentially lessen any excessive contingencies during preconstruction (Figure 4.3). In addition, BIM promotes best services, not necessarily best cost. Although an estimate may be checked through peer review or by an estimating consultant, the goal is to have the best number as opposed to a lower number.

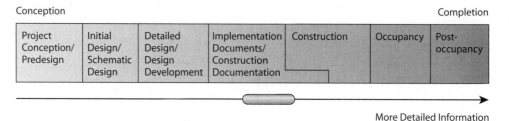

Conception   Completion

| Project Conception/ Predesign | Initial Design/ Schematic Design | Detailed Design/ Design Development | Implementation Documents/ Construction Documentation | Construction | Occupancy | Post-occupancy |

More Detailed Information

Phase Bar

Figure 4.3 A BIM model during the prebid stage is useful for defining the scope and budget of a project.

Some strategies for using BIM during preconstruction to help lessen contingencies are as follows:

- Get the subcontracting team involved early in the development of the design and project coordination. This increases the subcontractor's understanding and comfort with the project, and many times will give the subcontractor insight about the type of team involved prior to construction.

- Include in the contracts the ability for the subcontractor to be reimbursed. This strategy allows the subcontractor to be compensated if for some reason they are not selected and the owner decides to take the project to bid. This strategy is useful in that it allows the subcontractor selected to gain an advantage to understanding scope, budget, and project issues more clearly than an outside subcontractor bidding on the project for the first time. In turn, this gives the subcontractor a vested interest in the project because of the amount of time put into the project during preconstruction to be competitive.

- Coordinate constructability reviews with the subcontractor prior to issuing final construction documents because this will help alleviate a potential increase in difficulties envisioned through construction that the architect and general contractor may not have been aware.

- Use the model to assist in defining the scope of work and complexity of the project three dimensionally to the subcontractor.

The last point is what I will focus on in this chapter, because I've already covered early subcontractor involvement, which continues to make the case for accomplishing more integrated practice. So, how can BIM be used during preconstruction prior to bidding?

Prebid and Hard Bidding

If the project is a hard bid project and has not included any early involvement from the subcontractor, the BIM model is still useful but limited in effectiveness. By using Navisworks or similar software, the general contractor can quickly isolate items within a model that are specific to that particular subcontractor, create search sets or selection sets, and then issue the NWD file to the subcontractors for viewing. For example, if a mechanical engineer has completed the ductwork design in Revit MEP, the subcontractor can view this file using Navisworks if it has been layered and saved as an NWD file such as in Chapter 3's scheduling tutorial.

Although this is useful in the bidding stage, the model in this scenario is typically for supplemental information only and should not be the sole basis for a bid. Because the contract documents remain the primary means of establishing the scope and responsibilities for a particular subcontractor, often models aren't distributed among subcontractors bidding on the work. Many companies utilize a waiver associated with this model that allows the subcontractor to view the model, which is specifically used for reference only, and the contract documents still hold sway unless otherwise agreed to by contract.

How Do I Get the Models?

Many times after lectures and during BIM breakfasts, technology work groups, and training sessions, I get asked the question, "How do I get the architects' and the engineers' models?" Many times the stories go something like this:

"I asked the architect and engineer to send the model with the 2D drawings so I could use it, and they said no."

"I told the design team that they were issuing the 2D documentation, but they said they couldn't issue the model...why?"

continues

How Do I Get the Models? *(continued)*

In this book, I have talked about different project delivery methods, and the one in which there is the least amount of resistance among the design team to share the model with the general contractor or subcontractor by its very nature is an integrated one. So, how does it work in the design-bid-build world?

The answer, unfortunately, is that many times it doesn't.

Although there is a great opportunity for AE teams to get a lot of good clarification data and coordination questions by issuing the model, many firms won't issue the model to general contractors and subcontractors because of liability and contract concerns. This is where you can really see where old-school practice butts heads with new technology. When it comes down to it in a design-bid-build project, models contractually do not need to be issued to contractors bidding on a project using typical contract language. In addition, the contractor does not have a "right" to that data.

Now, where does that leave the contractor?

Well, the real issue here lies in the process of project delivery and the true value of BIM in regard to bidding a project. An IPD or advanced design-build project will typically have a much more open if not critical attitude of model sharing and potentially address BIM and file-transferring language as part of the process, whereas the value of BIM in a hard bid situation would be limited in that everything I have outlined so far would be difficult for a subcontractor or general contractor to realize in the amount of time allotted to bid a project. In fact, it might even spell disaster for the AE team to try to coordinate the model among multiple bidding general contractors and even more subcontractors. Overall, this type of communication has a greater chance of confusing everyone instead of helping because the sheer amount of information is often too much to digest in a four- to six-week bid schedule.

I'm not discouraging transferring models during a hard bid because they are extremely useful to experienced general contractors, but go into the request knowing that the architect or engineer might not have time to respond to model issues associated with a design-bid-build project. Model updating will be slim, if at all, because of time constraints. In a scenario such as this where old processes meet new technology, the outcome is often limited compared to more integrated forms of delivery.

Ultimately it's in the architect's best interest because better information equals better bids. So, let the design team and owner know exactly what you intend on using the model for and that you are willing to sign any disclaimers or waivers if necessary and hope for the best. Keep in mind, though, that unless you've established this process in the contract language, they are under no particular requirement to issue it to you.

Prebid and the Integrated Project

Budget updates present a unique set of challenges when they relate to a more integrated method of delivery. From the beginning, the selection of a subcontractor should be based not only on the capability of completing the construction of the project at a reasonable cost but on being able to meet the demand of a BIM process and supplying the required deliverables as outlined in the contracts. This coordination is extremely important in a BIM project. Team selection should be taken into consideration when project teams are being assembled. This selection is even more important when issues arise, such as when an owner has a "preferred" subcontractor they want to work with who is consistently low priced but lacks the ability to deliver a BIM fabrication model and an as-built BIM of the work after the project is completed as outlined in the contracts. In this example, communication with the subcontractor is key. The construction manager should make it known that the bid needs to include creating a BIM model and scope to coordinate it as well as part of the contracts. A construction manager in this instance will need to work with the subcontractor and make them aware of resources and options to meet the BIM criteria for the project. If the subcontractor isn't willing to utilize new technologies, the general contractor should make the owner aware of the subcontractor's inability and arrive at a solution between the team.

Another example might be a steel fabricator who is the low bidder on a design-build project and has the ability to perform the work but has a limited ability to deliver a detail 3D model for the record BIM, as was outlined in the BIM deliverable of the contract. These issues are the realities that BIM projects face, and they're better if addressed at the onset of a project or as soon as the issue arises. If these issues arise during a bidding process, there are often a number of feasible resolutions that won't involve the subcontractor raising the fee on a project beyond what it costs to have a consultant complete the models. In fact, a number of companies in the United States and abroad specialize in modeling services for construction projects both before and at the completion of a BIM project as mentioned earlier. In addition, if the fee on the use of a U.S. modeling company is too high, many subcontractors in the United States will outsource the modeling scope of a project overseas to meet the program requirements of a BIM project. Although this has been met with mixed results, it is still a task that needs to be managed, and additional fees might need to be figured into the cost of not only performing the work but of having someone manage the modeling for the project.

Engineers often have the ability to create a model for the subcontractor and might be able to provide an additional resource for a fee if the scope of work is relevant to the engineer. Of course, the ability for a subcontractor to hire an in-house BIM specialist is not always an option but is becoming more commonplace as the number of BIM-specific projects begin to rise in order to stay competitive. In this way, the subcontractor can estimate the approximate amount of time the specialist will be utilized on the project and introduce that fee as part of the project scope. Last, if the project is advanced and extremely complicated, some consultant companies specialize in applying their expertise to make sure that the project goals are delivered. Often this is a costly option but is feasible if the project is constructible but the means of modeling it is "over the contractor's heads." This option is also a good idea because it becomes the responsibility of the consultant to make sure the modeling is completed on time and that the contractor has a fixed consultant fee and is not spending internal time on a scope of work they're unfamiliar budgeting into the project fee.

KANSAS CITY ART INSTITUTE. IMAGE COURTESY OF MCCOWNGORDON. COPYRIGHT MICHAEL SPILLERS.

Sharing and Transferring Digital Information

Throughout the construction community, the practice of sharing and distributing CAD files to contractors and consultants alike has widely been accepted because general contractors, architects, and engineers have gained an understanding of what the files were being used for.

Digital file sharing among teams was met with skepticism initially because of a general lack of understanding about what file sharing could provide other design team members. However, as time has progressed, this has become more and more accepted, and in many companies it almost goes without saying that a general contractor or subcontractor has the ability to open and view CAD files.

Sharing and Transferring Digital Information *(continued)*

A discussion has arisen in the AEC community in regard to sharing and transferring BIM files and what liability and issues might arise—much like when CAD file sharing was an issue. The most common concerns are the following ones:

- Inadvertent editing of the BIM file

- Liability associated with interpreting the BIM as an "end-all" design tool

- Using the "legacy information" from a BIM to be copied and used in the future by other companies that might be in direct competition

You can solve most of these issues by using some of the best practices mentioned in this book. To limit accidental model editing, arrange for the proper contract documentation to be negotiated and signed at the front end of a project. If no project language has been established but BIM coordination is still a project goal, utilize other documentation such as a BIM waiver. A waiver clarifies the exchange and use similar to the information exchange and model coordination plan but focuses only on digital information sharing and model ownership. An example of resolution at a further stage might involve saving the file in a neutral file format such as Navisworks to protect legacy data, which allows users to still engage in the use of BIM and establish a comfort level with the team. Ultimately, it is important that the BIM is responsibly shared to increase the effectiveness of the team.

Sharing BIM files ideally requires work at the front end but can pay dividends when shared with a project team later. Especially when considering further integration of project teams and the technology age as a whole, digital file transferring will become critical to the success of tomorrow's successful companies.

Integrated teams also provide unique opportunities in using BIM when bid numbers on a project are close. Often a project is given to a subcontractor who has better qualifications because of previous BIM project experience and is able to contribute more to the design team than the other. Although selecting team members based on qualifications has become part of the process, BIM has begun to be factored into the equation more and more. During this phase, the construction manager really begins to see the contract language for file-transferring standards and formats tested. Prior to bidding (if applicable) and after construction or implementation documentation has been issued, it is often typical for more information in regard to quantities of updates, questions, addendums, responses, and clarifications to be issued than was issued to the general contractor to date. This barrage of late-coming information seems to always happen, so it is critical to maintain the project standards that were established at the onset of the project (Figure 4.4). As mentioned earlier, this is essential for a number of reasons, mostly in that the way you manage and track information now is reflected in how it will be transmitted to the field. In certain circumstances, someone will need to reformat the information to meet the standards prior to construction beginning on the project. Although maintaining standards seems like a feasible option, often it is not, because the tasks seem to increase only after a project has been bid, and document coordination begins to take a backseat to other more important tasks at this phase. For this reason, as well as holding the design team and yourself accountable to contractually meeting your obligation to the team, it is wise to take the time to verify that the information that is being transferred is being tracked and logged as required by the contracts and that the structure for the construction manager is being built correctly so that later during construction administration it doesn't fall apart.

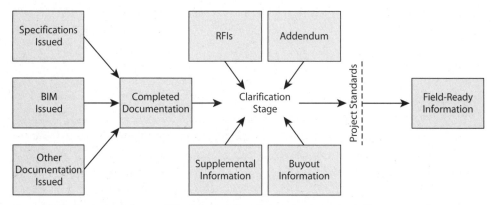

Figure 4.4 Clarification stage diagram. BIM reduces the influx of information during the clarification stage of a project.

BIM and Estimate Updates

Another strategy for leveraging BIM during a project is to use the BIM file for updating estimates very quickly. Later I will discuss how to use an updated model to quickly generate an outdated estimate without having to use significant resources and complete manual quantity takeoffs. It is important to note that last-minute design changes, addendums, clarifications, and scope alterations that are physically represented can be altered and updated much more quickly than typical takeoff methodologies can catch up with. By utilizing BIM technology during the estimating stage of a project, you can house the building documentation in a single model and use it to clarify many of the questions that might arise during the bidding process, from the schematic to the beginning of construction (Figure 4.5).

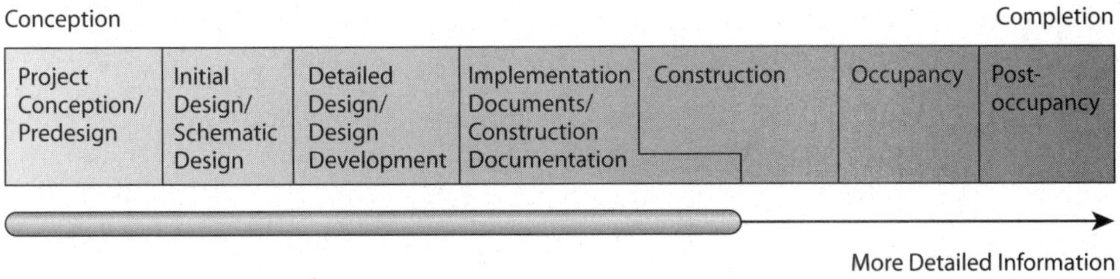

Conception   Completion

| Project Conception/ Predesign | Initial Design/ Schematic Design | Detailed Design/ Design Development | Implementation Documents/ Construction Documentation | Construction | Occupancy | Post-occupancy |

More Detailed Information

Phase Bar

Figure 4.5 Estimate updates are used from project initiation to construction.

Many times the "value of BIM" question comes up, and many times the answers seem to be in clash detection or improved visualization. However, one of the best uses for BIM is the ability to utilize the work already accomplished by architects and engineers and streamline the takeoff process. BIM and estimating can be simplified down to two common denominators that affect the quality and accuracy of the estimate:

- The quality and content of the BIM
- The quality and content of the cost database

In the tutorial to follow, I will demonstrate how to run an estimate update using Innovaya to update the quantities in your Timberline estimate as you refresh the link to the BIM. To provide detailed estimates, not everything needs to be modeled. Of course, every contractor would prefer virtually complete models, but because of current limitations in BIM file sizes, it is often a much better strategy to not model every nut and bolt but to instead communicate with the design team what the modeling strategy is. Using detail information such as detail components, callouts, text, and linked specifications is currently the hybridized method of using BIM. The hybridized approach to document creation is when an architect creates a model and then as the views of the model such as floor plans, elevations, and sections are created, there is additional 2D detailed information layered on top of this view (Figure 4.6). This enables the design team to still use the model effectively while not modeling down to a very deep level of detail and without significantly increasing file sizes. This is another area in which BIM will continue to develop as software continues to grow in sophistication. Regarding a composite model, the contractor should still use models from both the engineers and architects and layer any additional critical information onto the construction model or communicate with the design team where additional detail is needed (such as structural details) and where it isn't (such as in doorknobs).

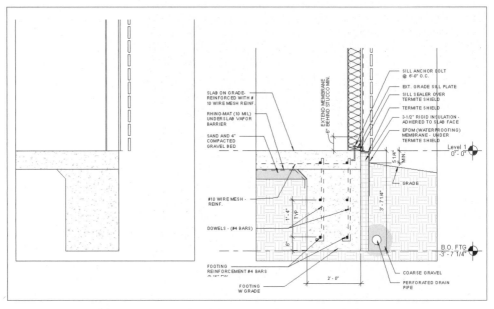

Figure 4.6 Example of the model view to the left and the same view with detail information on top of it to the right. The example image shows the difference between the default model view and the same view with detailing and annotations layered on top of the view.

So, if everything isn't modeled, how do you know exactly what you are estimating? The answer is that BIM provides a more accurate solution to the estimator in the form of quantity to assembly. Estimators are still very much needed in a BIM estimating practice. You can't click a button, and you're done with the estimate. The estimate and methodology of taking off still need to be verified for accuracy, and the estimate itself contains information that might change, such as square footage premiums and tight sites where the BIM model, if used as a stand-alone solution, would generate an update of quantities in the same format it was previously mapped to. What should be detailed, though, is the contractor's cost database.

Every contractor has a "typical" way of doing things, and many contractors see their typical way of doing things as a hybrid strategy. For some, this means that initial plans were made to use a certain piece of software, archive cost history models, and train new associates in a customized company tool. Often, what happens is that the least common denominator becomes the standard. The software that everyone is used to or able to be immediately used by new personnel brought on during "busy periods" becomes the estimating methodology of choice. BIM, set up right, can help transform an estimating department into a much more efficient and productive group.

To help visualize the concept of general modeling, imagine a 3′ 0″ by 7′ 0″ hollow, metal interior door. In Revit, for example, this component can be directly inserted into a wall, and its element properties can be filled out via text fields in a door schedule, including model, manufacturer, hardware type, and rating (Figure 4.7). Now knowing what you have learned through model mapping in earlier Innovaya tutorials, you know that this door can be assigned a cost assembly. These assemblies can include such items as labor, hardware type, finish, and so on. The question therefore becomes, does the architect need to model the door hardware for the estimate? The hinges? The closer? If they're in the estimate assembly, probably not. Unless you need some specialized or custom door, modeling to the "nth degree" in the example really isn't necessary. By using detailed cost assemblies and the concept of model component linking, Innovaya remembers the mapping of model components to cost assemblies even as the model changes. Thus, a drastic increase in productivity considering the model may use such tools as Auto-Takeoff. This Innovaya tool updates the cost almost instantly and identifies all unassigned items needing to be investigated and mapped to costs. Updating the budget is critical for project success. As such, it is important to use the latest version of the model to create the latest version of the estimate, as opposed to making assumptions and approximating these changes because of time constraints.

BIM cannot fix all the estimating and data management issues in the world, but it can be a valuable resource during the estimating or phase of a project. Using Timberline in this example, you will use the 75 percent design development model to run an estimate update. Then you will compare it with the previous 50 percent design development model.

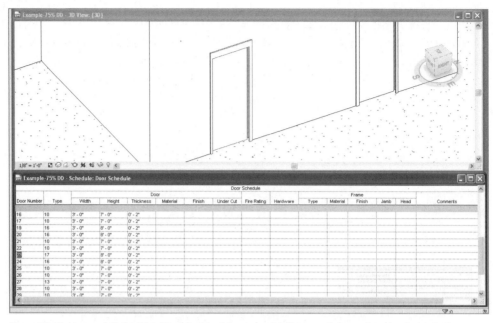

Figure 4.7 The door model's detail level is minimal but information rich. The example image shows how using the door schedule to host additional information to model components limits the amount of model detail.

Note: As you overlay the two files on top of each other using Navisworks later in this chapter, you will not use Innovaya's Merge or Synchronize model features.

To begin, it's usually a best practice to archive the previous Innovaya cost model and to create a copy of the Timberline estimate for future reference later in the project. This example utilizes an updated Timberline estimate from a new Innovaya file. Refer to Chapter 2 for reference in how to create an estimate. You can update an estimate from Innovaya in a couple of ways. The first is to use the copy mapping function from a previous estimate. This will copy all previously linked assemblies to the new model that can then be used to update the Timberline estimate. The second means is to archive the old model and save the new model file over the previous file. You can accomplish this by exporting the model from Revit to Innovaya Composer and use the previously linked filename.

Note: In general, when working with BIM, it is wise to keep the most current files saved as a general type for instances, "estimates", "schedules", "mechanical model", and so on. This will make archiving these files throughout the process easier when they can be saved as either dated files (`MEP model-4-30-09`, `MEP model-6-10-09`, and so on) or phases of the project (for example, `50 percent construction documents`, `75 percent construction documents`, and so on).

Comparing the New Estimate with the Old in Timberline

Timberline can compare side-by-side estimates and quantity takeoffs from Innovaya. In this tutorial, you will get a better picture of the implications of this new model update to the budget. Using Timberline, you will check for the variance between the estimates by running a variance report. In this example, I will show how to generate a variance report from the two estimates.

Creating a Variance Report in Timberline

1. Open Timberline.

2. In Timberline, open the Example-75% dd.pee estimate.

3. Now select Reports > Variance Report (Figure 4.8).

4. First select Example-75% dd.pee, and then click Add > New.

5. Select the completed Example-50% dd.pee estimate from the book's companion web page (www.wiley.com/bimandconstruction)

6. Click Add (Figure 4.9). This opens the Variance Report dialog box (Figure 4.10).

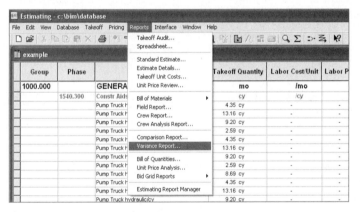

Figure 4.8 Using the Variance Report function in Timberline

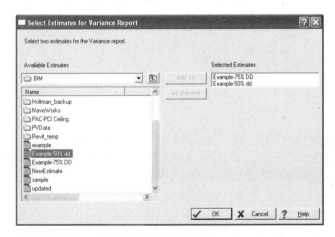

Figure 4.9 Adding an estimate to a variance report

Figure 4.10 The variance report options box.

7. Click Preview.

8. Click the Report Options button, which opens all the exporting and printing settings to export the report.

In this example, you will analyze only the quantities in this example.

9. Deselect all the fields under Report Options except for the Qty fields and the Total fields, as shown in Figure 4.11.

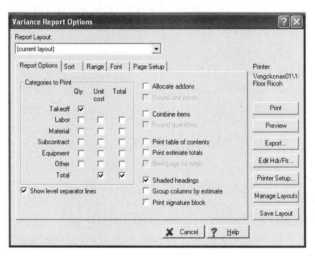

Figure 4.11 Using the Variance Report Options dialog box, you can customize the comparison to how you want to see it.

Exporting the Report to Excel from Timberline

1. Now click Export, which gives you the option to save the file as a number of different file types including PDF and XML. In this example, you'll export this file to an Excel Workbook file.

2. Enter the file location and name you want to save the report as, and specify Microsoft Excel Workbook (*xls) as the file format (Figure 4.12).

3. Open Excel and open the newly created report.

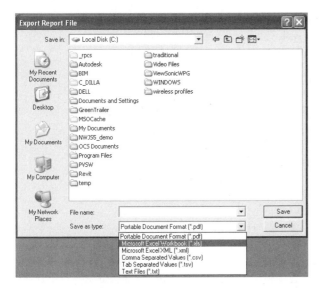

Figure 4.12 Exporting the variance report into Excel

You'll now see that the two estimates have been categorized with each other and changes have been identified in the Variance column within Excel. Additionally, you can now reformat this for printing and archive it for future reference.

BIM and Model Updates

Throughout the process of model sharing and transferring digital formats, you can update your current model in a number of ways (Figure 4.13). Depending upon the project delivery method type and type of company delivering it, you need a coherent and logical means of updating BIM information. The strategy for model updating in a company that has all design-build functions in-house or that colocates in the same office for a project will be different from a team in New York working on a project in Dubai, for example. Furthermore, the amount of information is limited in our digital world by the amount of data that current standards allow you to transfer over the Internet. I talk about future data transfer in Chapter 8, but in this chapter I'll talk about current suggestions for transferring data efficiently.

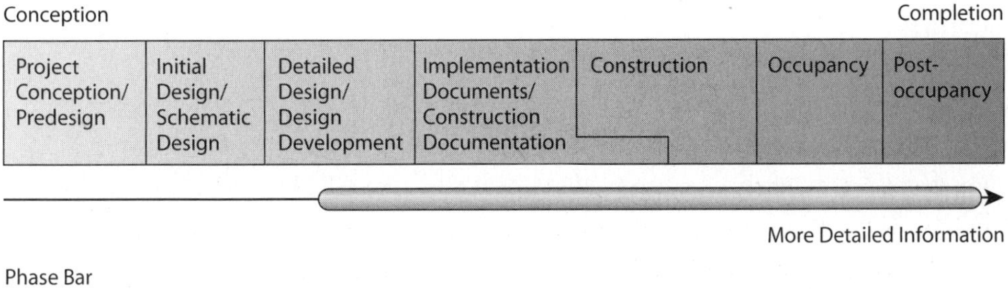

Conception   Completion

Project Conception/ Predesign	Initial Design/ Schematic Design	Detailed Design/ Design Development	Implementation Documents/ Construction Documentation	Construction	Occupancy	Post-occupancy

More Detailed Information

Phase Bar

Figure 4.13 Keeping the model current updated begins in design and lasts throughout the life of the project.

Building information models are only as useful to the team as they are current. Many times I'm asked, "Why do we always seem to be playing 'catch up' to the design team?" Or conversely, "Why can't the design team catch up?" The answer is old processes require information to be delivered to the rest of the team when a particular phase of the project has been completed. For example, the architect submits her 50 percent construction documents because she believes it is completed to this level. The problem for this type of data transfer strategy is that you are always looking at old information. The architect is not going to stop drawing and stop tweaking and modifying things as the general contractor reviews it. And an engineer is not going to stop working on a set of drawings after submitting the design development drawings. In this digital age, where information is critical to the success of so many decisions, some project teams parallel project information as a blog and updates as an RSS feed or live feed of information for a project. All team members want to base decisions and strategies on the latest data. Typically, if the data or drawings change, the decisions change. So, how do you transfer BIM files that contain a huge amount of data (typical BIM files are between 50GB and 200GB some can reach up to 400GB and 600GB!) back and forth between all parties?

If your operations are all in-house or are colocated, then you have the advantage of using network connections and can house the models on a networked server so that they can then be accessed, modified, and saved by the entire team, granting some exclusiveness to attempts to edit same files or components. The ability to save and pull down new information is more streamlined because the users' computers contain the software relevant to their work, not the servers. Essentially, the server acts as a warehouse, receiving and shipping data but not producing any of the information. Of course, this is a best-case scenario, where huge amounts of design and construction data can flow smoothly across the network and where, conversely, teams can receive real-time BIM updates as the project progresses. In addition, this is why a number of companies have adopted this strategy because it is the most effective way to work on a BIM project. What about the other companies that need to update their models as frequently as possible?

If you're like so many other construction companies, then you are used to CDs or DVDs of 2D file information that is couriered back and forth among the teams. This approach has really become antiquated; unless there is particularly sensitive security measures that require this type of transfer, it is one of the most ineffective means of delivering model updates. However, many companies now use an FTP or extranet solution, and in the following tutorial, I will show you recommendations for best-practice model updating. The first will show how to archive old models and link in the new files. The second tutorial will show how to use a received BIM model to update an animation.

The FTP model (file transfer over the Internet) holds the most promise because it requires only a high-speed connection to the Internet and can be accessed by all parties including consultants, owners, and local approval organizations. Although this is not a real-time means of viewing the model in its native format across multiple disciplines, it offers much better security and ease of access. (Refer to Chapter 2 for a more detailed outline of model-transferring standards.)

Updating a Revit File

Updating the Revit file is something that is required to keep the information in the model current. As mentioned earlier, it is usually a good idea to develop a system of archiving, as well as a file-naming convention, so that as a project progresses, you can create archives.

Updating the BIM Using Autodesk Revit

1. Open the Construction.rvt Revit model.

2. Open the visibility graphics window by either typing **VG** or selecting View > Visibility/Graphics (Figure 4.14).

3. In the Visibility Graphics window, select the Revit Links tab. This is where you can manage the links associated with the Revit file. In this example, you will merely turn off the Example-50% dd.rvt link and link in the new Example 75% dd.rvt file link.

Figure 4.14 Setting up visibility graphics

4. Deselect the link Example-50% dd.rvt (Figure 4.15).

5. Click OK.

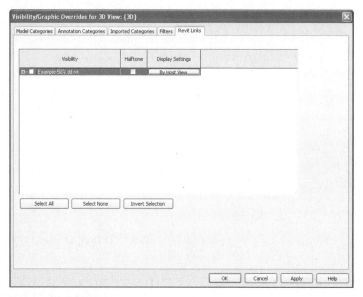

Figure 4.15 Unlinking the 50 percent version of the Revit file

You should now be looking at nothing but the layered floor you created in the earlier tutorial.

6. To continue, select File > Import Link > Revit.

7. Select the Example-75% dd.rvt file, and click Open.

Note: You can also click the pull-down bar located to the right of the Open button to import specific work sets. This is useful when large files are being handled and you need to see only specific components in the BIM model.

8. When Example-75% dd.rvt is imported, save the file.

This will link the latest version of the architectural model into the Construction .rvt file.

For reference sake, you can turn on both the previous version and the updated version and use the Half-tone, Transparency, or Color Fill settings to see what has been altered from the previous model.

This completes this tutorial. Keep in mind that new models don't necessarily need to be deleted from the composite model file; in fact, in early design stages, it's easier to leave them on so that as models go through multiple iterations, turning them on and off in one model is possible.

Updating a Navisworks Animation

This tutorial shows how to update schedule animations. Keep in mind as information, files, and schedules are linked into Navisworks, the software searches for the correct location of these files. To limit confusion, I will show how to do this as opposed to providing all the files required.

Saving the New Models for Use in Autodesk Navisworks

1. Open the file construction.nwf, which contains all the models and schedules for the project.

2. Save the file.

    Notice that this model is the older 50 percent model, which is linked to the schedule (Figure 4.16). In this tutorial, you will update the arch-model.nwd file with the 75 percent design development Revit model.

3. Open the arch-model.nwd file, and save it as 50% arch-model.

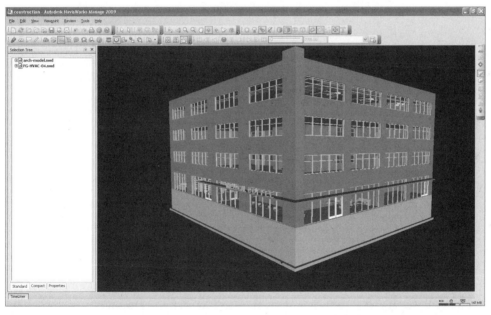

Figure 4.16 The existing Navisworks file prior to updating

Now you will export the Example-75% dd.rvt file from Revit to Navisworks.

Exporting the New Revit Model to Update the Navisworks Animation

1. Open Revit, and select File > Tools > External Tools > Navisworks 2009.

2. Specify the filename as arch-model.nwc.

3. Click Save.

Updating the Sequencing Animation Using the New Model in Navisworks

1. Open the new `arch-model.nwc` file in Navisworks.

2. Save the file, overwriting the old `arch-model.nwd` file.

3. Open the `construction.nwf` file; the new file should have replaced the older file (Figure 4.17).

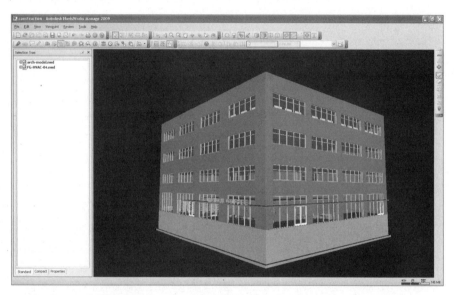

Figure 4.17 New linked version of the architectural model

The mapping to the schedule remains the same; any new items that haven't been linked to a task should be visible on the Timeliner Simulate tab at the beginning of the project.

You can update the schedule similarly by archiving the previous schedule and saving the new schedule over the current schedule. If the schedule is linked to a Microsoft Project schedule, then the file will update when saved and Navisworks is opened. The schedule must be exported if it was exported before. For instance, the example schedule export would need to go through the same process again to "repath" the schedule. Keep in mind that Navisworks identifies task names to linked model components. That said, if the name of the schedule changes or if new line items are added, they will abandon previously linked associations.

Note: Typically it is best to leave the Navisworks file saved as a general type of file, such as "construction" or "current", throughout the preconstruction and construction phases of a project and then save the old files as an archived version, such as 50% dd, 75% cd, and so on. This keeps the files backed up for reference if needed and makes importing new files easier, because they simply need to be saved in the same location with the same file-name as previously opened.

This updates the schedule link as well as the new model file. To add tasks that might have been added to the schedule, right-click the schedule link, and select Rebuild Task Hierarchy from All Links. This adds any new tasks to the project that weren't there before for additional model linking.

As shown, updating an animation sequence is straightforward, but it is important to note in Navisworks that in order for the process to work correctly, you must have a file archiving structure in place to be effective. This allows for future models and schedules to be saved over the old ones after they are archived.

Clash Detection Updates

Earlier I discussed clash detection and its value to a construction management team; in this section, I'll investigate its ability to be assigned to the responsible parties, updated, and then new clashes assigned. The value of BIM as a multiple trade coordination tool increases incrementally every time clashes are found, tracked, and resolved before a project reaches the field. This clash detection resolution/reporting allows construction managers to utilize BIM as an organic means of finding issues with those models provided by engineers and subcontractors. Although industry metrics tend to range from an average savings of 40 to 50 percent of field change orders, much of the savings is relatively difficult to quantify; therefore, the following argument arises: "If those issues would have been found by an estimator or project manager before construction, does that subtract from your calculations?" As arguments about BIM on either side of the table request metrics, it has become widely accepted that the ability for a computer to find issues in 3D is much more detailed, quicker, and accurate than light tables or CAD overlays. This being said, in regard to the value of BIM in clash updates, the process must continue to be smooth as it moves forward. Otherwise, the technology doesn't allow for a better process and ultimately becomes useless. So, how do you continue to use clash detection reports as a tool throughout a project?

Clash detection takes time. Be aware that in highly complex work, the initial clash detection reports can be somewhat daunting. The numbers of clashes may easily reach into the thousands, and it takes time to find out what is clashing with what, prior to putting the responsibility in someone else's court. In other words, this is a process change. Although the old way of doing things allotted for an on-the-go constructability review with some overlays and overhead or below-floor coordination, clash detection takes more time. I'm sure some people have mastered this, but as a generalization, this process does not become "quicker" with updates, especially when there are multiple stakeholders inserting models, such as the three models shown in Figure 4.18. This is the opposite of what you learned about cost estimating and the ability to update estimates and quantities much faster than before. Instead, you now find a process that takes time and requires efforts that weren't accounted for before. The user's efficiency will begin to decrease the amount of time it takes to perform the clash detection against various model components with time, and the use of search sets and selection sets increases the efficiency of clash detection analysis and reporting. Still, the fundamental analysis of testing structural models vs. mechanical models and structural vs. site concrete, and so on, takes time. The construction manager should plan for this in a BIM project. The amount of time varies by the complexity of the project and the experience of personnel, but the average time increases by 20 to 30 percent for BIM clash detection and coordination, compared to previous constructability reviews.

Figure 4.18 Three different piping and duct models from different stakeholders linked together. The added degree of complexity with multiple stakeholders and multiple models increases the amount of time spent to coordinate the project during preconstruction and construction.

One of the major advantages to using Navisworks is its ability to export the clash detection reports in various formats. Language formats such as HTML and XML allow for other programs to receive the reports and to use them easily. Even more recently, new software is using the BIM outputs such as the clash detection reporting feature of Navisworks to coordinate the information in a single data source. Using the Internet, XML, and server space, Constructware by Autodesk has become a useful tool during a BIM process in the preconstruction phase of a project. Constructware is construction management software similar to Prolog and CMiC and a host of other software systems that provide budget, cost and procurement, document management, and hosted (website) collaboration. What is unique about Constructware is its ability to track and manage clash detection reports through the site. This added functionality takes a lot of the legwork out of creating clash detection reports manually. By using the export to XML format functionality, clash detection reports can be uploaded directly into the interface and users or clients can log in and see what clashes there are and who is responsible for them.

If you don't have access to construction management software, you can handle clash detection updates in a number of ways. For project teams using an intranet or extranet, the exportable file might be an HTML file. This file type can be viewed over the Web and is available to the entire design team and can be saved to an archive for future reference through Navisworks and Freedom Viewer. The disadvantages to this format is its inability to be directly edited or responded to, especially in Freedom Viewer, which doesn't allow for commenting in either the HTML import or the XML import type of file review. Mostly teams use the HTML tool because of its visualization characteristics with X,Y,Z reference points to be shared among those using Navisworks Freedom.

If both the construction manager and the subcontractors happen to have licenses of Navisworks, then files can be imported and reviewed, and new clash tests can be added and distributed among the teams using the XML format. Using the XML export and import tool from Navisworks, users can import the clashes into an NWF file that can be viewed among the team using Navisworks. The following tutorial will show you how to create batch clash reports and import them into Navisworks. When completing the tutorial, keep in my mind how you might use this when you are communicating back and forth with subcontractors on a daily basis. Navisworks also does an excellent job of not duplicating clash geometries when merging geometries. By bringing the clash detection report into the database via XML, Navisworks really gives you the most flexibility in regard to not only Navisworks files but DWF files as well.

For the following sample tutorial, you will use the NWD file you created in Chapter 3. You will then analyze further the clash detection report that was previously generated, and then run a clash detection on an updated file. In the previous clash batch, you can see that the majority of the clashes are because of the ceiling and the ductwork interfering with each other on the top floor (Figure 4.19).

Figure 4.19 The ceiling is interfering with the ductwork.

To verify this, move into a view on the top floor, and highlight the item called 2´ × 4´ ACT System.

Testing a Model Assumption Using Autodesk Navisworks

1. Open Navisworks, and open the file Example-50% dd-linked.nwd.

2. Right-click the ceiling, and select Item Hidden (Figure 4.20).

   This hides the ceiling.

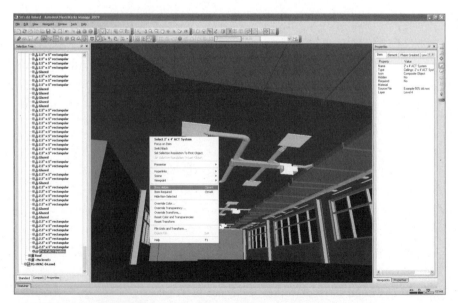

Figure 4.20 Hiding the ceiling element to limit clashes

Now you will rerun the clash detection report to verify that you were correct in your assumptions. This should reduce the number of clashes from 751.

Testing the Model for Clashes by Hiding Model Components

1. Open Clash Detective.

2. Highlight the FG-HVAC-04.nwd file in the left window and the Example-50%dd.nwd in the right.

3. Click Start to rerun the clash detection.

You now see that you were correct and that the ceiling accounted for about 533 of the clashes in the report.

In this situation, you now need to notify the design team that you have an issue with your floor to ceiling heights and that you need to either increase the height within the structure or lower the ceiling. Because the ceiling is at 9′ 6″, you might suggest both measures—lowering the ceiling to 9′ 0″ and raising the structure 1′ 6″ for a total shift of 2′ 0″.

Typically, this type of major clash would involve an email to everyone to identify the issue and try to propose a solution quickly, so the next revised clash detection report is somewhat manageable. Additionally, this issue could be tracked using web hosted project management software as well. Whichever is chosen, the issue should be brought to light quickly, distributed, and tracked.

To continue this tutorial, you will export the report and send the report as well as a viewable reference file in some form to review the clashes in. The problem is that neither of the exportable types from Navisworks is directly editable, and they require some work to get them into an editable format that allows the file to be commented on.

One way to do this is to export the file into an XML form, which can then be imported back into Navisworks on another user's machine for review and approval as well as into other programs, such as Excel or Constructware.

Creating an Updated Clash Report Using Autodesk Navisworks

1. Run the clash detection, and go to the Report tab in the Clash Detective window.

2. Select XML.

3. Select Write Report.

4. To import the XML clash report, select File > Import > Clash Tests XML (Figure 4.21).

5. Specify the file you want to import, and then click OK.

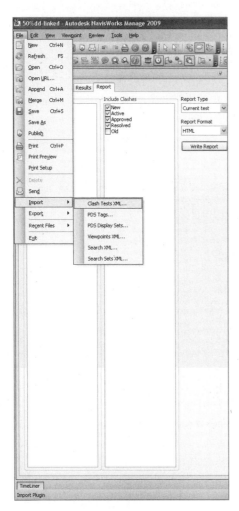

Figure 4.21 Importing a clash detection report

Note: To open this file in Microsoft Excel, find the XML file, and when the Open XML dialog box opens, select As a Read Only Workbook, and click OK. This opens the file in Excel. Although this isn't an immediately useful tool with macros and scripts, the XML to Excel path can become a usable tool. I won't cover Constructware in this book, but just know that you can import the clash report into Constructware as well and distribute it for delegating and reviewing responsibilities.

Once the XML file has been created, you can send it to another user who has Navisworks who can import it for review. Keep in mind that Freedom Viewer will not import XML clash detection reports.

This imports a clash report from another user into your Navisworks file, allows you to see specifically what the other user is looking at, and lets you get more detailed information about what is in each person's scope. Ideally, the workflow would look something like this:

1.   The report is generated.

2.   Fake clashes or minor clashes are marked or resolved.

3.   The remaining items and responsibilities become an XML file, which is distributed to the team.

4.   The respective models are altered.

5.   The new files are sent to the model manager, old files are archived, and new files are loaded into Navisworks for the next clash report.

It is usually a good idea to print or create a clash responsibility matrix in a spreadsheet and distribute it at the same time as the clash report. Distributing during progress meetings or as the clashes need to be resolved lets users identify which issues are theirs and which ones aren't.

This seems like a feasible solution, but in reality all parties involved might not have licenses of the Navisworks software.

Another means of exporting the clash report is as an HTML file. This method allows you to view the exported clash report through Explorer, and the HTML file can be imported into Adobe Acrobat Professional for commenting.

Exporting the Clash Detection Report to Adobe Acrobat Professional

1.   Run the clash detection,

2.   Go to the Report tab in the Clash Detective window.

3.   Select HTML.

4.   Select Write Report.

5.   Create a separate temporary folder because large amounts of images are associated with this file.

6.   Save the export.

7.   Open the folder the report was just exported to.

8.   Open the HTML file.

9.   Click Save As, and select Webpage, Complete (*.htm,*.html). Save the file in the same directory (Figure 4.22).

10. Open Adobe Acrobat Professional.

11. Click Open, and select the HTML file just saved.

12. Save the file.

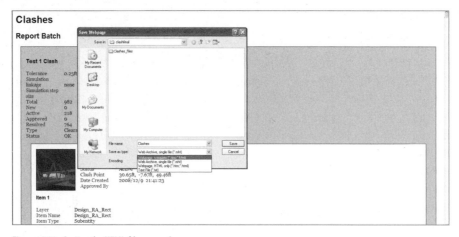

Figure 4.22 Saving the HTML file as a web page

Now that your clash detection report is in Adobe Acrobat Professional, you can set up responsibilities and commenting in a number of ways. The easiest way to enable commenting is to have the issuer identify whether the clash is critical. The user sending this report can create a comment on the clash instance to identify whether this is a fake or minor clash as well as comment on what two parties have an issue through the commenting tool (Figure 4.23). This enables the mechanical engineer, for example, to search the Adobe document for comments containing the word *mechanical* in them, which will pull up all comments containing that word. Additionally, you can customize Adobe documents with forms that attach to the PDF file and audio files linked to comments as well to meet a variety of user's needs. A good reference for Adobe Acrobat Professional customization is *The Pocket Book of Adobe Acrobat 8 Professional* by Andy Zhang.

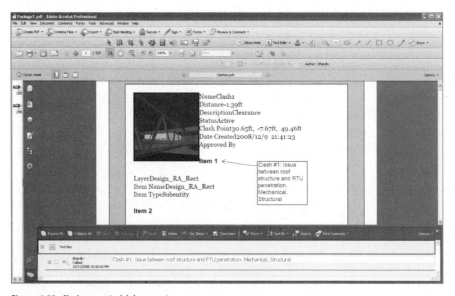

Figure 4.23 Clash report in Adobe to review

To conclude this portion of the tutorial, you will link this file to Adobe Review Tracker. This program allows a document to be stored on a server and uses an RSS feed to email the other users when the file has been reviewed and updated.

Setting Up the Clash Report for Review Using Adobe Review Tracker

1. Open the Adobe file.
2. Click the Review and Comment icon.
3. Select Send for Shared Review (Figure 4.24).
4. Walk through the step-by-step guide. Completion of this tutorial enables users to work in the same file provided it's on a read/write network folder.

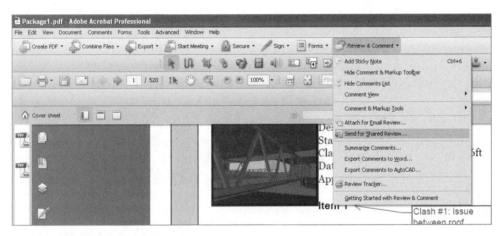

Figure 4.24 Setting up Review Tracker in Adobe

Updating the Clash Report

You will now use the earlier tutorial's clash detection file to update the sequencing animation to update your latest Navisworks files. If you haven't yet completed this, please refer to the earlier tutorial titled "Updating a Navisworks Animation." This example shows you how to rerun your clash detection to see whether the ceiling issue has been resolved. In this example, you are simulating that you received an updated model from the architect and told the mechanical engineer to compensate for a shift of 2′ higher elevation from level 1 and up for every floor. This will allow the architect to move the existing ductwork model to the new Z coordination. For reference, open the new Example-75% dd.rvt file in Revit, and move to an elevation or section view where you can see the new plenum space design for level 4, which is 3′ and is in between elevation 52′ 2″ and 55′ 2″ (Figure 4.25).

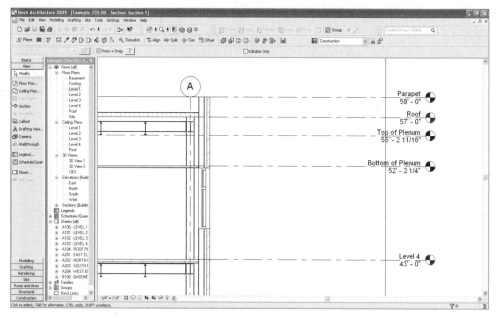

Figure 4.25 Revised plenum space in Revit

In this example, you will use the latest models as if they have been exported from Revit and are ready to use and open in Navisworks. Then you will run the clash detection.

Updating a Clash Detection Report Using Autodesk Navisworks

1. Open the file Construction.nwf.

2. Select the arch-model file on the left screen and the updated mechanical model mech-model on the right (Figure 4.26).

3. Leave the defaults, and click Start.

You now have created a new clash detection report. The new file should have somewhere around 107 clashes (Figure 4.27) with no large-scale issues on single components. This means that the construction manager can begin digging into the clash report and finding out the details of what is conflicting with what and whose responsibility it is. Ideally, you will use Navisworks in this scenario so other users can find the clashes they are responsible for and can update their models for additional testing. A clash detection responsibility matrix lets you more easily identify issues, responsibilities, and the changes made to the model.

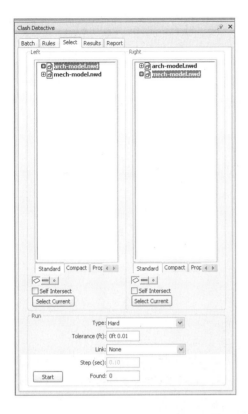

Figure 4.26 Updating the clash report with two new linked files

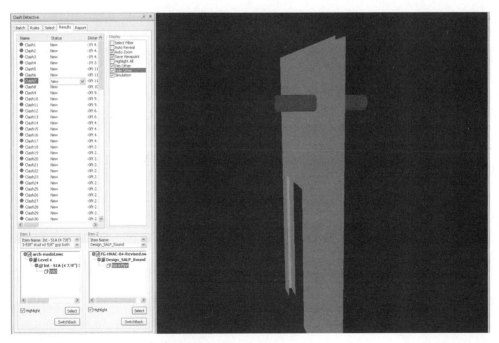

Figure 4.27 Revised clash detection ready for more coordination

This tutorial just scratches the surface of how to use a clash detection tool, but it is good to know how the information is able to flow from all parties and be compiled in a truly collaborative piece of software for all parties to test and analyze (see Figure 4.28). Navisworks is a unique tool in that it has the ability for teams to update the information as it is transmitted and then directly imported into the software. Other software does allow for updates and for different file formats to be imported and exported, but the industry needs more software similar to Navisworks that allows for teams to better collaborate without colocating. The discussion for interoperable standards will continue to be defined by formats that the most people are asking for and that all software can work with. That may be the software and processes that you're familiar with, but it might not be. Navisworks is one of the first interoperable standard in that it allows a way for virtually all file formats to be imported into a composite model; it also allows further collaboration by exporting into formats that can be imported into other software as well.

IMAGE COURTESY OF BRAD HARDIN.

Figure 4.28 Live clash detection meeting to update and review a clash detection report with the team

BIM and Budget Management

This section of the chapter discusses the potential opportunities for BIM and how it's used in the budget management process. By updating the model throughout the prebidding process, you are able to analyze the model and produce results in the form of clash detection reports, sequencing clash reports, model updates, and archiving. In regard to bidding, a more integrated approach to design and construction would not have any portion of the project go out to bid unless it was a requirement set forth by the owner. This would mean that there isn't necessarily a bid day but rather a final spot check of the GMP, verification of subcontractor's scope, and material pricing. However, in the older, traditional approach of project delivery, during this stage the project is "sent out to bid." This process involves sending out great amounts of information to many subcontractors who are then responsible for supplying bids in a relatively short period of time.

In regard to BIM, more integration is better that traditional methods during bidding and budget management for the following reasons:

- Better pricing does not mean a better project. Because the subcontractors' number is low and the bid would make the owner happy does not mean that the subcontractor who has had no involvement with the project until now has any desire to focus on the project's success rather than his own, and it opens the general contractor up for potential "nickel and dime" change orders that arise from the subcontractor as they try to collect on their lost fee due to "incomplete project documentation."

- The process of putting the subcontractor stakeholders who were involved in a project earlier and not awarding the project to them based purely on cost weakens relationships unless there is a buyout agreement in place. Regardless, there is a trend within the industry to complete "bid checking" with subcontractors' competitors. Yet in a process where those competitors have not been actively engaged in a project and are not aware of all the issues and constructability constraints, these numbers might vary greatly because of a fundamental lack of understanding on a project.

- The myth that early involvement of a subcontractor equals a lower price is just that, a myth. Although the advantages to having the subcontractor on earlier in the process far outweighs the alternative, the belief that as the subcontractor gains understanding about the project his number is always going to go down is flawed. Although his contingency might be lower, as I discussed earlier, this does not necessarily reflect a lower number for the project but rather a *better* number for the project.

The opportunities for software companies moving forward is to streamline the process of not only prebid and bid management but also project management through the use of BIM. This means creating tools for the industry to use that provide for a single hosting solution for the general contractor to offer to subcontractors, providing answers to those who don't know how to use BIM through the ability to use 2D drawings, and more importantly embracing the future of data transfer beginning with the ability to collaborate through BIM software. This type of information tracking and commenting will provide more resources and information to team members and continue to further the cause for a more integrated approach to design and construction.

Conclusion

Using BIM as a tool to update the flow of information during a project is just as critical as doing it the first time. BIM tools will continue to develop in their ability to be streamlined, but current processes allow a BIM-enabled construction manager to complete tasks with relatively good efficiency.

As users become more proficient with the software and the process to which the software is applied, these tasks will be even faster. In this chapter, you took an updated Revit file and linked the new file into Innovaya to update your quantities and costs, you used the updated Revit file to run a schedule clash detection report in Navisworks, you exported your clash report to Navisworks and Adobe Acrobat Professional, and you updated the new models with a new clash detection report. Updating estimates and maintaining the constructability issues of a project are valuable components to a construction manager. These tutorials showed you how to use these software systems. You can begin to see how the process works as you move into the field.

BIM and Construction Administration

BIM can be a powerful tool during construction administration. Used in conjunction with a digital request for information (RFI) and submittal process, using BIM in the construction administration phase of a project provides the project superintendent and supporting staff with many tools that were not available before. In this chapter, you'll learn about the following topics:

In this chapter

Overview of BIM and construction administration

BIM and RFIs

Training the superintendent

Setting up the job trailer

BIM and field coordination

BIM and punch lists

Overview of BIM and Construction Administration

As the BIM industry continues to develop architectural and engineering modeling tools, BIM will continue to provide the contractor with more ways of finding and coordinating that information. Yet, as the industry continues to develop tools for the architectural and engineering teams, it is also critical that software companies learn how contractors are using building information modeling and what they would like to see more of. This chapter provides some insight about how building information modeling can currently be utilized and offers some suggestions about the areas in which software still needs to develop and begin bridging the gaps of a BIM process and project. Additionally, it should be stated that BIM in fact does "work" in the field and that it's often a misconception that although it's not completely intuitive and integrated that it doesn't function at all. There is plenty of room for improvement and smoother interoperability between field systems and tools; however, there is a great opportunity missed by contractors who choose not to utilize this technology and leverage it to some or its full extent on the construction site.

The model coordination plan continues to play a critical role in a BIM process as what was developed at the beginning of the project defines who uses the model, where the model gets distributed, and what it's used for during construction. In the field, trained superintendents with BIM experience may use BIM to do the following tasks:

- Produce digital RFIs
- Perform 4D scheduling updates
- Create a field sequencing clash detection
- Clarify the installation of fabricated components
- Add as-built and in-field model information
- Run field clash detection reporting
- Use the BIM for punch lists
- Prepare the model for project closeout

The model coordination plan should define the tools used during construction and the file formats that are acceptable to use during this phase in the project. The information exchange plan should outline how the information will be transferred and audited. For example, is the architect going to give the model to the contractor to use freely during construction, or does an approval process need to be implemented? Additionally, the information exchange plan should state who will maintain ownership of the model and provide the necessary changes throughout construction to the BIM files. These issues, among others, are unique to a BIM project, where the transfer of data and the means and standards of exchanging data are of particular importance. As more BIM projects become case studies and as these issues are resolved with more standardized regularity, best practices and lessons learned will emerge. As stated earlier, it is simply impossible for all the questions that arise in a project to be planned for and outlined in the model coordination plan. For this reason, some issues will need to be resolved by the team as the project moves forward with a mind-set of flexibility.

Existing Process

To begin, let's analyze the typical flow of information as it pertains to a 2D or more typical CAD project. In a CAD project, after the 100 percent construction documentation and addendum and revision information have been turned over to the contractor, construction begins. Before the first scoop of dirt is moved, it is the contractor's responsibility to verify that the immediate information needed is adequate and the most recent. Further issues for the project need to be identified and put on a path of resolution for all scopes of work. In theory, this review period for the project allows the construction manager to identify issues and give responsibility to the correct subcontractor(s). In itself, this is an arduous process. Analyzing and overlaying CAD files or sheet drawings is time-consuming work, it lacks in visibility, and it is prone to errors and missing information. In reality, the construction manager often can't juggle managing the project documentation, the trades, the field management, and the construction manager's own management team. Especially in a less integrated project approach, such as design-bid-build, the architectural and engineering teams often seem to take less interest in continuing to document the project to standards, and therefore the project's documentation turns into a mishmash collection of CAD files, PDFs, faxes, and scans. Although this sort of documentation may be fine for a team that may build smaller projects in an "on-the-go" methodology, for more complex projects this type of documentation processes could be the bane of a project's success. Ultimately, the amount of staff needed to correctly review these documents may require a significant investment. For this reason, many field personnel rely on fixing issues in the field because of a lack of adequate documentation. Construction management teams use software such as Prolog, CMiC, Constructware, and others to track and issue RFIs because the process of reviewing the information, attaching data and drawings, and forwarding them to the appropriate team member or verifying that the team member resolved the issue with the latest documents is time-consuming and because the software helps to alleviate and automate some of these tasks.

In a 2D type of project, three things primarily stand out as inefficient and costly in regard to the time it takes to accomplish them:

- Overlay and document coordination review
- Changes in information
- Unlinked project administration tools to documentation

The first is overlay and document coordination review. In a BIM project, a significant advantage to using BIM as an overlay tool is that light tables and even referenced CAD data goes by the wayside. By using real X,Y,Z coordination points, the models can be viewed in all dimensions, not just two. As shown in the tutorials in Chapter 4, model updating involves archiving the old file, saving the new file as the old file's previous name, and reloading it into the composite model. In CAD, new sheets or drawings need to be issued and overlaid again on a light table or as an external-reference. Conversely, in-field BIM coordination relies heavily on the accuracy of the model when it is modeled as it is to be built. This means that the architects, engineers, and contractors who are creating models need to be sure that the level of accuracy in their model is to a level of constructible detail as the tool is carried into the field for coordination.

Edit a Dimension? But Why?

One of the most interesting discussions I have been part of was at a national CAD user conference, and at the end of a presentation on BIM, the speaker was asked, "Drawing to real accuracy doesn't allow you to edit dimensions and dimension strings, though! If I have a 3 $5/8$" stud wall and I dimension to the face of the stud, I want to round up to the nearest inch. Why would anyone want to model all of their dimensions if exact BIM doesn't allow for dimension editing?"

The speaker calmly responded by asking the attendee where he believed those dimensions in a string of walls went that he rounded up to. The attendee then responded that they were just included in the contractor's tolerances and was convinced that the contractor wouldn't construct something to the nearest $3/8$". The speaker responded by saying "First, if I have a string of 10 walls and all of my $3/8$" dimensions are rounded up, that then means the last wall will be a total of 3 $3/4$" short. Now if this is an ADA hallway, required clearance, or, worse, the boss's office, I don't want to have to explain to him why his office is smaller than the other's because he was on the end!" The speaker went on to explain the importance of measuring exactly as you want the structure constructed and that avoiding coordination issues was much more important than "clean-looking, completely inaccurate dimensions."

The second issue that's inefficient in a 2D project is managing changes within new information. In CAD, the medium of transferring information is either in the form of updated sheet drawings and CAD files and is usually done via email or in the form of documents transferred via FTP or other transfer software. The issue with transferring CAD data is that, in many cases, archiving the files becomes increasingly complex, as does the readability of plans, because multiple CAD files become layered upon each other from different iterations and can become cumbersome. Traditionally, during construction, the old files are turned off, hidden, or deleted from a centralized CAD file. Because there are so many layers of information for a construction team to review, the focus becomes just coordinating the latest information as opposed to comparing design changes with previous files. This leaves a large gap in coordination capability because other trades might have based decisions on previous files as well as budgetary changes that might have an impact on the total cost of the project.

Lastly, with 2D projects, project administration tools are not linked to documentation. This issue is of particular relevance and differs from BIM significantly. For example, in a CAD process, there might be six separate CAD drawings to resolve a ductwork coordination issue, and all six drawings are issued to the engineer and construction manager. Although these same six areas of resolution need to be shown via six separate CAD drawings, in a BIM process, resolving all six issues is accomplished by updating one model. This results in one 3D model being issued for coordination, testing, and approval, using the software to do the hard work as opposed to using six different drawings, which requires manual coordination and review. Additionally, single CAD drawings may contradict one another because of a lack of coordination. Checking a single CAD drawing against a host of other CAD files that might or might not affect the new file is neither the most efficient nor effective means of document coordination. In many instances, BIM will reduce the frequency with which construction management tools are used to track information, questions, and documentation. In BIM a construction manager will need to identify which model addressed certain questions or which new model will address other issues or changes that arise. In essence, Band-Aid documentation during construction administration goes by the wayside in a BIM project because multiple issues are resolved in one model and then issued to the team for review.

New Process

In BIM, as everything is modeled and files are compiled into a composite model, previous models can more easily be compared to new models. Design changes can be tested against other trades for clash detection, sequencing, or deviation reporting more efficiently. Using BIM effectively means that rarely will there be just a portion or piece of a model submitted for review; rather, one revised model will be resubmitted. Because BIM deals with "completed" models, this further eliminates coordination issues between numerous incongruous files. However, similar to CAD, the process to update, review, and test these models is a time-intensive effort. This is not because of sorting through the information to try to identify issues but rather because of coordinating all of the issues found by the analysis tools. Whereas before, setting up the CAD files and layering the information took a significant amount of time, now with BIM, the time is spent reviewing issues found between the revised models. For this reason, BIM is a much more effective means of coordinating information and multiple trades by using the software to find issues than wasting time trying to identify them manually. Because BIM requires a similar amount of time to coordinate model updates, it also requires teams to have standards in place to archive old files and issue new reports based on testing tools to the team in the field as well.

As I discussed in earlier chapters, the need to address issues such as how BIM will be utilized to issue RFIs, manage in-field clash detection, and keep a detailed 4D schedule must be outlined during the preliminary contract negotiation. The plans agreed upon must be carried forward to actual practice in order to succeed. Using BIM effectively is not something that can be done in a haphazard manner and must be thoroughly thought through. Additionally, collaboration in a BIM process does not stop in the contractor's office at the end of preconstruction. In fact, BIM now becomes a more valuable tool in the field where the work is occurring, and by stressing the importance of construction accuracy, the project's partners will begin to have confidence built into the project. Projects will never be perfect; there will always be some unforeseen circumstance, and this is the very nature of the construction industry. This is a good thing because it means computers can't do it all and that the tool continues to be only as useful as the person who knows where to click! More important, it means that the coordination accomplished beforehand determines the amount of coordination in the field.

BIM and RFIs

Building information modeling and requests for information (RFIs) are somewhat of a new science. In the realm of architecture and engineering, BIM has begun to take a foothold in the industry, and the processes related to BIM are beginning to be shaped. In the construction sector, however, BIM's value during construction to closeout remains somewhat ambiguous (Figure 5.1).

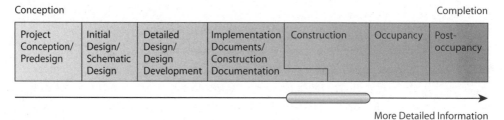

Figure 5.1 RFIs begin prior to construction to project closeout

This is because of several reasons; here are two:

- The first reason is that the educational curve among contractors is just beginning to grow. By "educational curve," I mean that the actual discourse of lessons learned and best practices within the contracting community are beginning to be defined by user groups, conferences, and word of mouth. However, even these meetings rarely get into the inner workings of BIM during a project, and it is left to the general contractor to define processes and move forward.

- Second, new technologies are often met with some degree of reservation. This is not because the tools don't work, but because no one wants to be the first to dive in and find out if it's shallow water. Although the purpose of this book is to help in defining a process, each contractor works differently and has unique software and processes that define the way they work.

So, where do you start in defining the processes necessary? To begin, you define the goals you want to achieve through BIM, and then you can move toward achieving them. Of course, one of the main goals is better project coordination, but more specifically the ability to house the information in a usable file with easier team access might be a goal. That said, let's take a look at how a BIM-based RFI can take place.

To begin, we will look at Adobe Acrobat Professional. Adobe software is widely used in many professions as well as in the construction industry. The advantages to Acrobat are that Acrobat Reader is free and PDF is widely accepted as a format in which to send, mark up, and receive documents. PDFs allow multiple documents to be combined and sorted in a single format in a read only format if desired. Adobe Acrobat Professional offers the ability to combine both 3D information and 2D sheet drawings. Conceptually, this means that all the needed information is housed within a single file; however, this is not a linked set of documents. The model is separate from the sheet drawings in that if the model needs to be updated, the change will not be made throughout all of the documents. This means that a new 3D PDF or sheet will then need to be created, and the old one will need to be deleted or archived.

As discussed earlier, the Review Tracker tool essentially acts similar to how an RSS feed would on the Internet. When the file has any new updates or changes, Review Tracker sends an update notification to the entire team through the feed. This may be either an email with a link attached or a downloadable plug-in that, once installed, allows the user to see what, when, and where information or items have changed within the project. This can be extremely useful to teams working on larger projects with tablet PCs in the field as updates are automated, tracked, and sent to the team. Acrobat also includes digital signatures via the Stamp tool, which cuts down on the amount of printing, signing, stamping, and faxing required and instead allows architects and engineers to review the files and use a digital stamp that verifies whether the item is approved or needs further review. Adobe has also added functionality that allows users to virtually meet through a webcast. This functionality allows users to mark up, highlight, and comment on the Acrobat file in real time through their computers. The webcast meeting functionality in Acrobat is similar to an online meeting type of webcast software but is file type based and is unique in that it allows the document to be shared and edited, not just displayed. The tool that you will use in this tutorial is the Review and Comment tool. This tool is customizable and tracks in the same file who created the question, when it was reviewed, and the response. If the team is using Adobe Review Tracker, the update will automatically be sent any time the file is saved.

Currently, Acrobat allows both parametric models and 2D documents to be contained in the same file. This is unique to BIM review software, with the exception of some plug-in software that houses a fully linked file or the use of two separate files, a model file and a sheet file, such as in Autodesk Design Review. Many contractors seek solutions for a more robust linked set of information that maintains both the latest BIM with the current sheet data and specification information. Although Adobe offers a solution, there is certainly more room for growth in toolsets, usability, interoperability and linking capabilities. Regardless, for this exercise, you will create an RFI PDF sheet from Revit. Once the sheets are created, this tutorial shows how to merge the two files together.

 Note: New model views created by the construction manager for RFI clarifications are typically better in quality and resolution than scanned files. However, it is best to review with the team any phases, worksets, or model components that may be turned off or hidden prior to working with a model.

To begin, let's create the RFI sheets in Autodesk Revit.

Creating a Sheet RFI in Revit

1. Open an .rvt file.
2. When the file is opened, create a new sheet to put the new RFI view on.
3. Select View > New > Sheet (Figure 5.2).

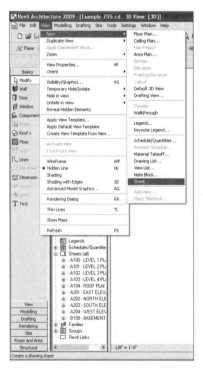

Figure 5.2 Creating a new sheet in Revit

4. In the dialog box, select the titleblock you want to load.
5. If 8.5 × 11 is not an option, select Load.
6. Select Imperial Library/Titleblocks.
7. Select the A 8.5 × 11 Vertical sheet (Figure 5.3), and select Open, and then click OK.

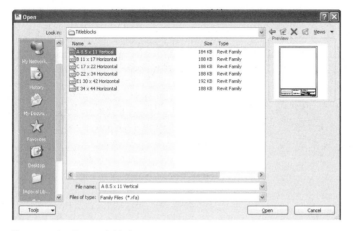

Figure 5.3 Loading a titleblock

8. Go to a view such as Floor Plans Level 1.

9. Create a new view by zooming into a detail such as the bathroom bank located at the center of the building.

10. Select View > Callout.

11. Verify the floor plan is selected in the drop-down menu.

12. Drag the window so that it encompasses the women's restroom, as shown in Figure 5.4.

13. Once you've created the callout window, highlight the crop box, right-click the box, and select Go to View.

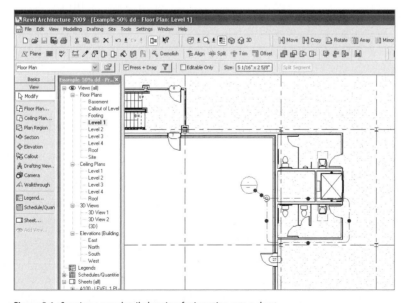

Figure 5.4 Creating a new detail plan view for insertion onto a sheet

You should now be looking at an enlarged plan of the view you just saw earlier. Now you will also notice that there is a new floor plan you created as a result of this callout. Since you selected that you wanted the callout to be another floor plan, it categorized the view under floor plan. If you had selected Detail View, the view would have shown up under the Detail View heading. Next, you will place the callout on your new 8.5 × 11 sheet by opening the sheet you just created.

 Note: The sheet should say *unnamed* and not have a plus sign to the left of it under the sheet heading, meaning that the sheet is empty.

Inserting the New View onto a Sheet

1. Once the new sheet is opened, select the Callout of Level 1 floor plan view, and drag and drop it onto the sheet (Figure 5.5).

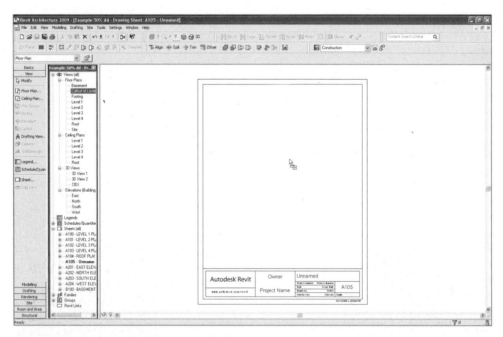

Figure 5.5 Dragging the new detail plan view onto the sheet

Now you'll print the new sheet as a PDF.

Printing the New Sheet as a PDF

1. Print the sheet as a PDF by selecting File > Print and selecting Adobe PDF as the name of the printer.
2. Click the Properties button.
3. Click the Layout tab, and select Portrait to orient the sheet correctly (Figure 5.6).

4. Click OK and then OK again, and then select where you want the new PDF to be saved.

5. Click Save to print a PDF; by default this will open the new Adobe file in a separate window.

Your new PDF should look something like Figure 5.7).

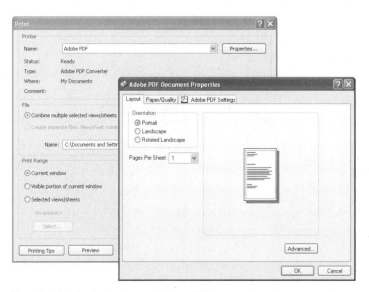

Figure 5.6 Printing the newly created sheet to PDF

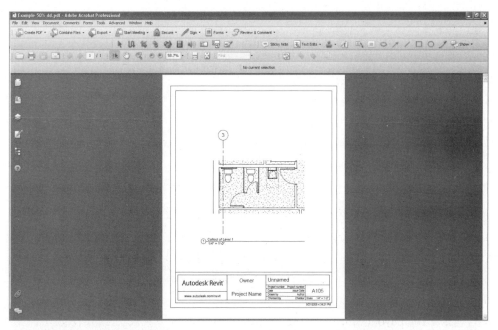

Figure 5.7 Example of PDF printed sheet

Within Acrobat Professional, you will now create a revision cloud markup. To do this, select the Cloud tool (Figure 5.8).

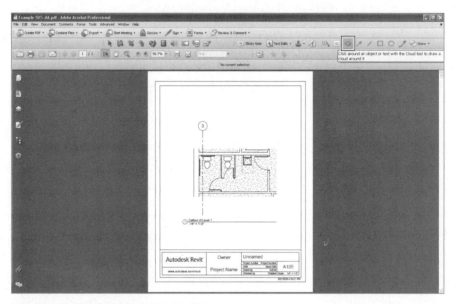

Figure 5.8 Using the Cloud tool in Adobe to highlight an issue

Commenting in Adobe

1. Draw a cloud around the door similar to Figure 5.9.
2. Select the revision cloud you just created.

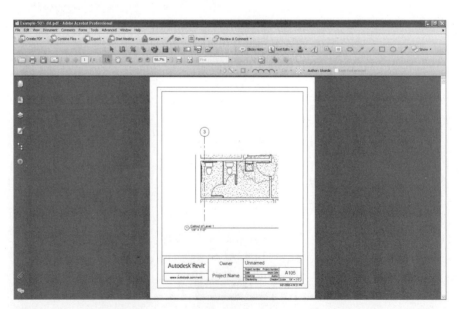

Figure 5.9 Creating a revision cloud

3. Right-click the cloud, and select Show Comments List.

4. On the dialog box at the bottom of the screen, click within the frame, and type **RFI 101 - What is the needed clearance from door swing to sink?** (Figure 5.10).

5. Save the file.

You have now created a sheet RFI using Revit and Adobe.

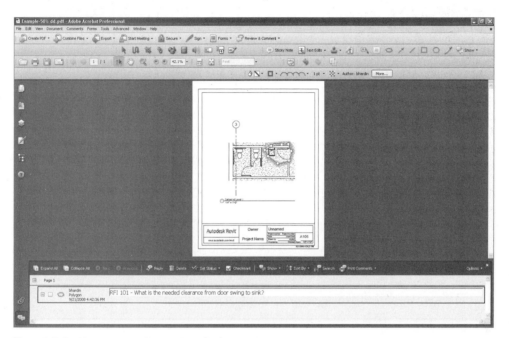

Figure 5.10 Entering text to complete a revision cloud comment

Using Review Features within Adobe

1. Using the RFI you just created, with the Comments dialog box showing on the bottom of the screen, click the comment you just created.

2. Click the Set Status drop-down menu.

3. In the drop-down menu, select Review (Figure 5.11).

4. This menu shows what the current status of the RFI is. This menu shows None, Accepted, Cancelled, Completed, and Rejected. To continue the tutorial, exit from the Review drop-down menu and instead click the Reply button (Figure 5.12).

5. When you click the Reply button, it enables a "sticky note" response. Here in this text box enter the text **12″ required clearance at this location from door to sink.**

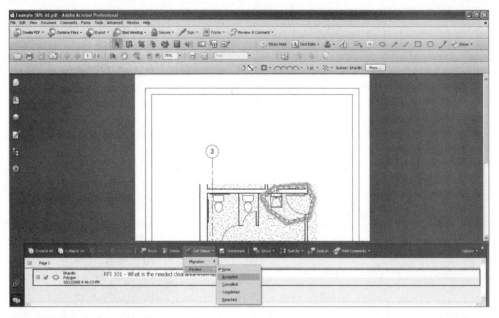

Figure 5.11 Using the Review drop-down menu in Acrobat

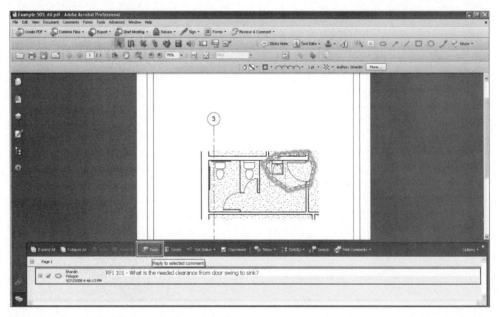

Figure 5.12 Using the Reply feature in Acrobat to respond to review comments

This could be the response for the RFI from the architect or responsible team member. In Chapter 4, I showed how to send a clash report for shared review; here I will expand on how this PDF can be sent out for review. The first tool that Adobe provides is under the Comments header and is Send for Email Review, which allows users to send the PDF via email. By using the Merge and Respond to Comments function within Acrobat, the document remains a single file that is housed on all the users' local machines, and the users are notified when comments, responses, and updates need to be merged into their file to update it.

The second update tool is the Send for Shared Review option. This option hosts the file on a network file or FTP site and retains the single file or files to be updated throughout the project. Users receive an email link to install Adobe Review Tracker. Similar to an RSS feed, the plug-in sends a query to the file, and when the file is updated, a notice is sent to the team. This option is useful for a number of reasons; it limits the amount of files the users need to have on their hard drives, and it works better for web-enabled coordination. The Review Tracker tool is proprietary to Adobe and needs to be installed separately from Adobe itself.

The third tool is the Upload for Browser Review function. This function notifies users through email; however, the attached URL link sends the users to the location of the file on the network. This allows the users to still maintain a central single file that is shared by all users but doesn't require any downloading of the Adobe Review Tracker software.

All of these tools can be extremely useful depending on your needs and approach. Additionally, all of the review-tracking functions in Acrobat walk the users through a step-by-step process of setting up the file or files to be shared with the project team. Adobe allows users to not only post graphic clarifications but to post text-only Comments. Another great feature of Acrobat Professional is the ease of customization of forms, templates, and digital stamps. All of these make for a great tool, with the exception that it lacks the functionality of linking sheets to a BIM file. The ability to update the model file and then have it filter through Adobe's robust tracking software would make Acrobat a powerful RFI tool, yet currently this functionality isn't offered.

Training the Superintendent

Superintendent training should be relevant, focused, and to the point. A superintendent or foreman will typically not need to know how to create linked estimates, create sequencing animations, or create BIM contract language. However, the superintendent will need to know how to create in-field RFIs, improve coordination in the field, update the model, and use the BIM for punch lists and project closeout. In addition, superintendents might learn how to generate a logistics plan or run a clash detection in the field. Overall, it is typically the way the company or firm is structured and how it desires to run its operations that will determine how much and to what level of detail the field personnel on a BIM project are expected to accomplish and be trained.

The ideal timing for training a superintendent is when the superintendent completes a course in BIM and is immediately followed by a project to which it applies. When this method of training is not possible, it is best to schedule BIM training prior to the BIM project beginning. An improper assumption made too often in the industry is that once an associate is trained, that associate will retain all the information from the course whenever the next BIM project comes up. Instead, all too often the superintendent will have intermediary projects in which he falls back into accomplishing tasks through the old process. Then when a BIM project does start, the superintendent is met with a level of frustration because he has forgotten the BIM training he received.

Generally speaking, training field personnel is somewhat different from training office personnel, and the levels of education and understanding of software can vary widely. The challenge in providing training for new BIM software and new processes to field personnel is gauging the level of associates. If possible, you should group new users with associates who have already completed training or have prior experience and can support the other. Many times this methodology is most effective when a senior associate has a younger associate who can help with the new processes and programs alongside the senior staff member. In turn, the senior associate can provide experience and insight into in-field project management to assist the other associate. Many times contractors have an in-house BIM team that can assist remotely or visit job sites to answer questions and provide support. Either way, field associates must have the following to successfully leverage BIM:

- A support system (either in house or external)
- A basic understanding of the software
- An in-depth understanding of the new processes as defined by their organization
- An understanding of how BIM software will make the project more coordinated
- Knowledge of where they will save time and what tasks will take more time
- An understanding that on BIM projects "typical" processes won't work

Superintendents and field personnel typically retain the most knowledge with in-the-field training. This connects the dots from the software and helps them do the tasks they need to accomplish (Figure 5.13).

To implement a solid foundation for training field personnel, begin with an introductory course into modeling software such as Revit, ArchiCAD, and so on. You can often find courses through your local software vendor, training consultants, or user conference; these courses are useful for outlining the basic functionality of the applications. This should then be followed by incrementally more advanced levels of training in two other areas: modeling and analysis tools.

Figure 5.13 Superintendents using BIM in the field to coordinate construction—applying learned technology to the field

Note: Many times the first group to complete a high level of training becomes a resource to the company so that there is no need to engage outside resources for training.

Training goals for basic skills Make sure that the course outlines basic modeling skills, such as how to create walls, doors, ceilings, and so on. Also verify that the course outlines how to edit these components. Remember that the purpose of training the superintendent is to gain an understanding of how BIM is different from CAD or paper drawings, to learn how they can drill down to find needed information within the model, and to understand how they can use the model as a functioning piece of building information in the field. The ability to become informed about an issue prior to issuing an RFI or other clarification can potentially limit the amount of RFIs issued on a project as well as add a superior level of visualization to the project.

Prior to project implementation, a basic level of training about new contracts, responsibilities, and BIM related goals should occur. This will be become particularly invaluable later in project application, when it must be made clear to field personnel whether the BIM is to be viewed as supplementary information that is absent from the construction documentation or whether it's to be used as modeled to be constructed. If BIM language is not in the contracts, the BIM model is usually for "reference information only" and still becomes an invaluable tool for the superintendent to use in communication with an architect or engineer. If BIM language is in the contracts, then BIM can be leveraged for a number of tools during construction, as outlined in this chapter. Additionally, if the model plan requires the model to be altered to represent "as-built" conditions a higher level of training and experience will probably be necessary.

Advanced training goals to create models More advanced training in model creation software is critical to manage the evolution of the fabrication BIM, to the construction in-field BIM, and lastly the *record model*. That said, it is important for the field personnel receiving training to gain an in-depth understanding of the modeling software and an advanced level of training in how to integrate and what to expect from other team members. The current reality is that too often fragmented, unreferenced, and inaccurate data is distributed between the construction team and then handed over to the owner to be used as information for the maintenance of the facility. The creation of the record BIM from the construction BIM must be completed as accurately as possible to the standards being requested by the owner and facility manager. If it is not the intent of a contractor to train their associates to such a high level of modeling skill, there are a number of strategies for delivering a record model that might influence the decision to train field personnel more in-depth:

- One option is for the architect to maintain a record model and update the model as the superintendent sends the AE team updated information. The format for this information can be typed documents, sketches, faxes, or models. The architect then reviews all documents and updates the record model. The model provided by the superintendent is often the more useful one because altered sections can be overlaid with model linking so the architect can see changes to the building and update the model. The advantages to this are that the architect (who might be more familiar with the modeling software) is able to maintain and track an accurate representation of the facility as the project nears completion. The disadvantage to this type of process is that often an architect will not have enough construction administration time budgeted to have the staff update the model. Although this should typically be a change in BIM process, sometimes this is not a feasible option.

- The second methodology for a model to be updated in the field is for the superintendent to update the model and send the model to the architect for approval throughout the construction process (Figure 5.14). While this requires a higher level of training, this is a good way of increasing efficiencies while still maintaining the accuracy of the BIM file, which the field personnel are using to construct a building.

- Other times a separate construction manager, owner's representative, or virtual construction consultant might be present in a project and have the ability to edit the construction model and provide a record BIM.

Figure 5.14 Superintendent using Revit in the field to update aggregate information into a record BIM

For the most part, field personnel utilize modeling software to verify dimensions, locate fabricated components, update the record BIM, and visually communicate to subcontractors a complex or three-dimensional reference. Other software can provide dimensional references; however, the tools, snapping, and interface are typically very accurate in the modeling software. It is best to outline at the beginning of a project who will be responsible for using and editing the construction BIM through this software, which will eventually become the record BIM.

Training courses to analyze BIM files The last area of training field personnel will be how to analyze BIM files. This training should inform associates how to maneuver in analysis software, test BIM files, and create a composite model. More specifically, the training should show field teams how to create RFIs, manage clash detections, and edit schedules.

RFI software as it relates to BIM should be able to generate two-dimensional and three-dimensional content from the composite model. Selecting the right software for creating BIM RFIs should be specific to the construction company and should dovetail if possible into existing software and processes. Numerous kinds of software can generate BIM-enabled RFIs, such as Acrobat, with Acrobat Professional edition allowing users to create 3D models that can be commented on. Other software such as Autodesk's Design Reviewer uses the model file to inform the project stakeholders what some of the issues are and it provides a free viewer and is a separate tool from the modeling software. In fact, these programs do not allow the user to virtually model anything, but rather they function as viewers that reduce the file size from a complete model to file sizes that are usable by superintendents and foremen who might not be using high-powered machines. Smaller "viewer" files are good for beginning personnel to use who are worried about involuntarily altering the model, because the model can't be altered. Currently, the BIM RFI process is somewhat of a disconnected one in that if a user desires pinpoint accuracy, full functionality, and views that are all linked in the same file, they most likely will utilize the modeling software (Revit, ArchiCAD, and so on) or Navisworks. However, if a user desires to use the RFI and logging functionality, they will use a BIM RFI production type of software, and there is limited interoperability between the two.

The other analysis tool training, such as clash detection and scheduling updates, should focus on what tools the firm deems valuable. There are no right or wrong answers to the level of training field personnel might receive and often the same team members from the preconstruction end of the project will continue into the construction phase. For this reason, it is a highly specialized and unique aspect of how a construction company chooses to operate. I have seen teams who have the BIM department run all aspects of a project in the field and I have seen teams where field staff are updating the budget via updated BIM files. Both were using BIM how they best saw fit and as a result were reaping the rewards in the field.

Tutorial on 3D RFIs

The following tutorial utilizes Adobe Acrobat Professional to combine the sheet drawings and a 3D model for a project into a single file for field utilization. Note that having a single file is sometimes not possible in Acrobat because the number of sheets and the size of the information require multiple PDF files to effectively coordinate information. Additionally, at the time of this publication, the ability to combine sheet drawing information and the model into a single file for editing and review does not exist within Autodesk Design Review; neither does the ability to colocate the sheet files for a project within the model into Navisworks. However, Navisworks does support flat 2D CAD information to be inserted into the model for further coordination. Additionally, only Design Review has the ability to insert sheet comments back into Autodesk Revit for further coordination.

Clearly interoperability between *all* of these systems is still an area in which BIM tools can be further improved upon. Two of the current popular tools are Acrobat and Navisworks. There are two schools of thought on the role of sheet drawings in this discussion:

- They are needed as reference for liability and accuracy concerns.
- Since the sheets themselves were derived from the 3D model and housed in the same file, sheet drawings aren't needed to use for coordination in the field.

It has been my experience that currently *both* types of information are needed. As discussed earlier, there is a certain degree of detail and additional line information layered on top of simple model views; conversely it would be difficult to navigate the information using the model only. Ideally, in the future, there will be a hybrid file management strategy that could then be phased from the VDC BIM into the record BIM. However, current documentation strategies require coordination in both a modeling tool and additional detail information as well.

The current strategy for keeping a record BIM is difficult throughout construction because of a lack of tools that communicate back to the native Revit file. Design Review works for sheets, and Navisworks' Switchback feature is useful for changing the model based on issues found in Navisworks; however, neither incorporates full functionality for both tasks. In this tutorial, I will show how to export an IFC format model from Revit into Adobe. There are other file types and other means of doing so, but in this example I will use IFC.

Creating a 3D Adobe File for Field Review

1. Open the file Construction.PDF from the website (www.wiley.com/go/bimandconstruction) in Adobe Acrobat Pro Extended.
2. Once the file is opened, go to the blank separate sheet called 3D overall view. Once you're at this sheet, you will create a 3D view.
3. Open a file in Revit.
4. Go to the 3D view in the file.
5. Select File > Export > IFC (Figure 5.15).

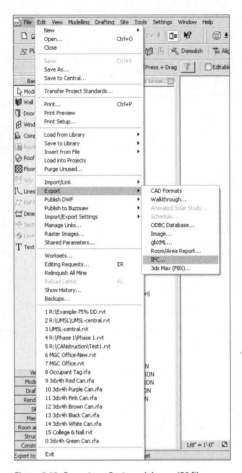

Figure 5.15 Exporting a Revit model as an IFC file

6. Save the IFC file.

7. Switch back to in Adobe Acrobat Pro Extended, and select the 3D tool from the icon bar (Figure 5.16).

8. Now in the blank area on the sheet, click and drag a box in which to place the 3D model (Figure 5.17).

9. When that is complete, an Insert 3D window pops up (Figure 5.18).

10. Specify the IFC file, use the defaults, and click OK.

Figure 5.16 Using the 3D tool in Adobe to import the IFC file

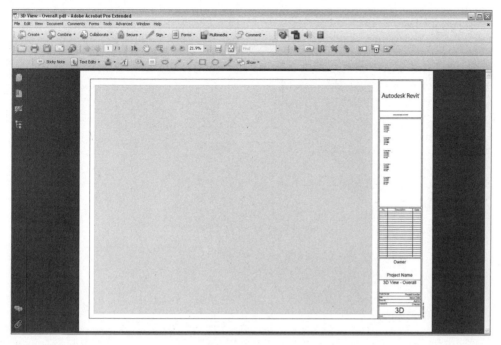

Figure 5.17 Selecting the area window in which to host a 3D file

Figure 5.18 The Insert 3D window specifies where the model is located.

This will take a moment, but once imported, you should have a file that looks like Figure 5.19.

To activate the window, click the box, and a new icon bar with navigation and markup icons will appear.

As you can see in Figure 5.20, the floor has been highlighted, and the intelligence associated with that BIM component is now displayed. Additionally, by using the 3D Comment tool, you can create comments and questions on the 3D file that track with the rest of the comments from the sheet drawings (Figure 5.21).

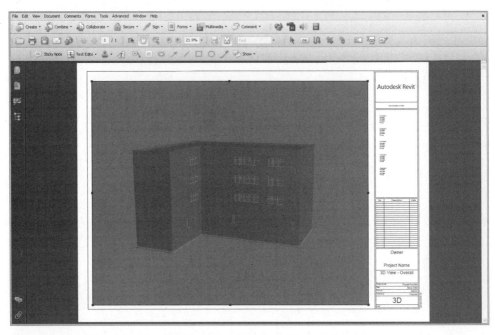

Figure 5.19 Example of a successfully imported parametric model into Adobe for 3D reviewing

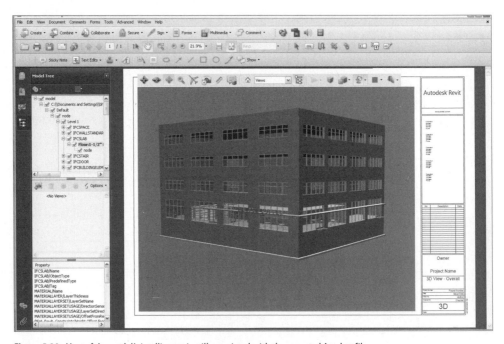

Figure 5.20 Most of the model's intelligence is still associated with the exported Acrobat file.

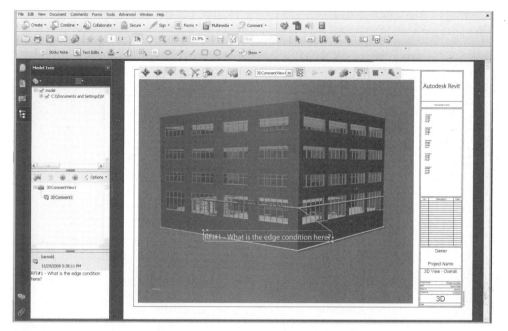

Figure 5.21 Comments on the model create related views and can be tracked and logged just like sheet comments can.

Navisworks Tutorial on Commenting

This tutorial will focus on utilizing Navisworks as an alternative in-field construction administration tool. Navisworks in general is capable of producing RFIs, but the cost of Navisworks for RFIs vs. the cost of other RFI-producing software usually makes Navisworks less competitive. Navisworks provides the superintendent with tools that allow for changes and alterations to be checked against the existing model in the field. This process is a new job function to construction management, and although this can be accomplished in the construction office as well as in the field, it is a best practice to train associates on tasks that connect with what they do. Developing an in-house standard of how to handle in-field clash detections and issue coordination will need to be sorted out prior to the beginning of construction. Often the superintendent and field engineer, who are the ones most connected to the project and on the job site the majority of the time, are the best to start training with.

In Navisworks, these are the three tools for commenting:

- Comments: These are associated with clashes, saved views, selections sets, tasks, and animations.

- Redlining: This is annotation that can be added over a viewpoint.

- Redline tags: These are tags used for recording issues during a review, and they combine the functionality of comments, redlines, and viewpoints.

Comments

For this example, you will use the Comments tool first.

Commenting in Autodesk Navisworks

1. Open a Navisworks file, and select a view you want to use.

2. Save the viewpoint by selecting Viewpoint > Saved Viewpoint > Save Viewpoint (Figure 5.22).

3. Now click the Comments icon (Figure 5.23).

4. Dock the Comments window in place, right-click a blank space in the window, and select Add Comment (Figure 5.24).

5. In the pop-up box, type a comment or question you have, and click OK.

Figure 5.22 Saving a viewpoint

Figure 5.23 Comments icon in Navisworks

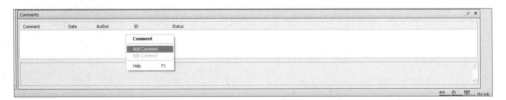

Figure 5.24 Adding a comment to the composite file

You will now notice that there is a new comment in the comment area, and it has tracked the author, the date and time it was created, and its current status. To edit the status of a comment, right-click the desired comment, and select Edit Comment (Figure 5.25). This opens the Edit Comment window where you can answer the question and change the status of the comment. Figure 5.26 shows the status changing from New to Active (Figure 5.26).

Figure 5.25 Editing an existing comment

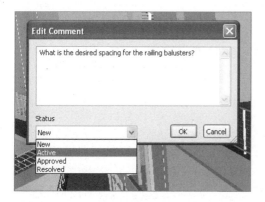

Figure 5.26 Comments on the model create related views and can be tracked and logged just as sheet comments.

Redlining

Next you will use the Redlining tool.

Creating a Redline in Navisworks

1. Using the same view that you previously used, select the Redline tool icon, and dock the window into place (Figure 5.27).

2. In the dialog box, specify which type of redline you would like to use; in this example, use the Ellipse tool. Other tools include Text, Redline Tags, Freehand, Line, Line String, Cloud, and Erase.

3. Select the Ellipse tool, and highlight an area for commenting, as shown in Figure 5.28.

Figure 5.27 The Redline tools icon

Note: You may edit the thickness you want to use prior to using the redline tools on the Redline window under the thickness pull down. The default is 4″ and is often too heavy.

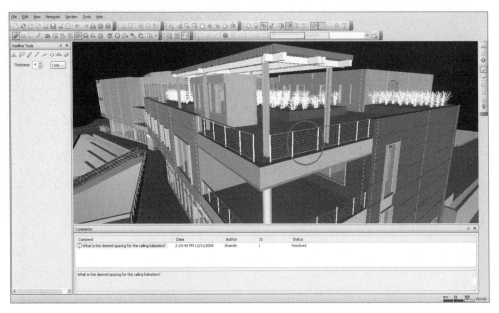

Figure 5.28 Using the redlining Ellipse tool to identify an area of concern

4. Now select the Text tool, and click the ellipse you just created.

5. In the Enter Redline Text field, type a comment, and click OK (Figure 5.29).

Figure 5.29 Adding a comment to the redline

Notice that the text entered shows up on the view, but it doesn't show up in the Comments window (Figure 5.30). Redlining is view specific, and a more common methodology of revision in Navisworks is the use of a Redline Tag.

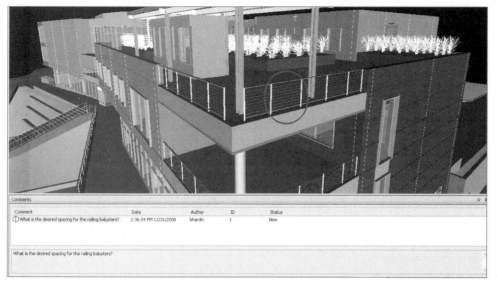

Figure 5.30 The text is separate from the comments.

Redline Tags

Redline Tags are probably the most common means of marking up a Navisworks file. A Redline Tag automates view creation and lets you comment on anything in the model.

Creating a Redline Tag in Navisworks

1. To begin, rotate the current view in the Navisworks model to a new view, but don't save the viewpoint.

2. Select two model components.

3. Now click the Tag tool (Figure 5.31).

4. Click once where you want the tag leader to begin, and click again where you want the callout to be.

5. This opens a pop-up window that allows you to add a comment (Figure 5.32). Insert a comment or question, and click OK.

Figure 5.31 Adding a Redline
Tag to model components

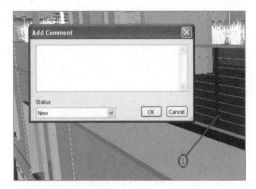

Figure 5.32 The Add Comment pop-up window

You will now notice that there is a new comment in the Comments window as well as a new viewpoint called Tag View 1.

You can also create Redline Tags on clash reports by highlighting the clash in the report window, clicking the Tag tool, and associating a comment with it. If done correctly, it should have a red dot in the Comments window (Figure 5.33).

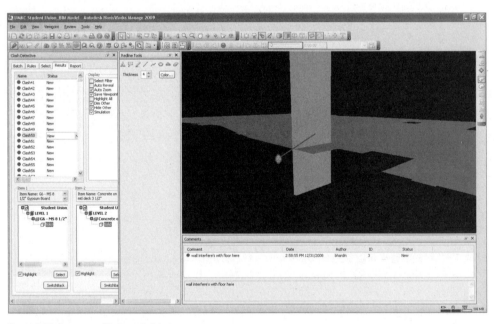

Figure 5.33 Comment added to a clash item

Additionally, the tags are numbered sequentially and can be searched by choosing Review > Comments > Tag.

BIM software will continue to evolve in the field of BIM and issue clarification. As the software develops, construction managers will be much more receptive to the software because they will have a fundamental understanding of existing software and practices and how it makes their work better. In essence, superintendent training is limited only by the user's capacity to understand the software and its functions. As noted earlier, schedule editing and clash detection training are other ways for the superintendent to use technology to verify trade coordination. The variance a construction project has from the 100 percent construction documents to what is built in the field is often significant. Although a BIM process aims to limit that variation, construction in itself is a complicated process and is almost impossible for an entire project to be built exactly as modeled across multiple trades and stakeholders. Thus, the value of a flexible process continues to rise in importance because the construction manager will be using the current BIM to coordinate changes throughout a project.

Setting Up the Job Trailer

Correctly setting up a job trailer at the onset of a project is critical to successfully using BIM in the field (Figure 5.34). Although this might seem like somewhat of a menial task, this portion of the process requires a new way of looking at how to display and share information on a job site. To begin, let's look at conventional job trailers; what they contain and what they are used for.

Generally speaking, construction trailers contain a conference room, a plans and specifications table, and a communication hub:

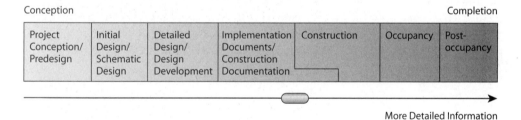

Phase Bar

Figure 5.34 Preparing the job trailer for successful use of BIM in the field begins at the beginning of construction.

Conference room The conference room functionality of the trailer is used for project team meetings, where all members of a project meet to discuss work completed, discuss issues, and work out scheduling. The conferencing function of the job trailer involves meetings among the managing members of the project who then delegate responsibilities and tasks to their workers to complete the work and keep the project on schedule. This round table, or *really long table* as it's sometimes called, involves a huddling of those members who then receive direction from the project superintendent about which work is to be accomplished that day or week. The superintendent relies heavily on the project manager to produce a working schedule that can be utilized by the team to define what work needs to be done to hit important project milestones.

Plans and specifications hub The second function of the job trailer is its ability to be as a plans and specifications hub. This role is due both to the physical location of the drawings and to the understanding that if the information isn't on the physical drawings, then someone in the trailer knows the most current information. With this in mind, the flow of information to the job trailer is usually directly filtered through a construction manager's office to the field, unless they are one and the same. Just as it is an absolute necessity to keep information in the trailer current, it is also a must to have a means of distributing that information to the subcontractors performing the work. This is typically done through faxes, email, or printed drawings that are sent to the subcontractor in meetings digitally and then delegated to the workers to perform the tasks. Distribution of this data and how it is relevant to the parties responsible keeps project management members continually busy. Although construction management software provides methods of transferring, logging, and sharing files, there are usually multiple single documents that are generated in response to a specific question. This documentation, although centralized, is usually segmented and not shared with all members of the project, partly because of the amount of irrelevant data that each subcontractor would receive but more so because of its format.

Using plans and printed schedules to communicate among the teams as well as specifications is critical to a construction team. All of this information is typically in a printed format and, if updated, contains "cut and paste" spec sections and drawings to which the construction team can refer. The superintendent is typically responsible for updating the drawings in the field through printing, cutting, and pasting what he receives from the construction manager's office via email or through document management software. Note that the direction and the flow of information at this stage in the construction process is oriented toward the field, because the project has reached a stage where the architect and engineer play a backseat to those who are performing the work.

This fundamental shift in information flow and focus is the product of the project shifting emphasis from design to construction. Thus, there is a shift in responsibility from the AE team to the contractors constructing the structure, based on the design drawings. The ability for a subcontractor to construct the design relies heavily on the contractor's ability to interpret the drawing information correctly and convert it into constructed work. If you assume that this is a typical design-bid-build and not a design-build or an IPD type of project, then you can also assume that the contractor is now using the information in the job trailer to build from. Although the subcontractor might have seen the drawings earlier during the bidding process, it does not necessarily mean the subcontractor has seen the drawings at all if there is a separate estimating department within that company. That being said, the need for current drawings in the field is invaluable.

Communication hub The last function of the job trailer is its role as the communication hub of a project (Figure 5.35). If there is a question or if issues arise, the first place these issues are brought up is in the trailer. Primarily, decision making on a construction site is based on the superintendent's experience with similar projects, the subcontractors' experience, drawing information, specification information and input from the AE team. Although there can be many other factors, in its simplest state the quality of the documentation is often one of the main reasons for site conflicts. Accurate interpretation of the information is critical to a project, but often differing drawing assumptions lead to potential issues. If the issue cannot be resolved among the field personnel or the subcontractor, or if the information is not clear, then an RFI is issued.

Many times disparate drawing information from multiple stakeholders is the cause for field issues as well. An example of this would be a mechanical engineer who responded to an RFI issued by the HVAC subcontractor. The construction manager then forgot to copy the electrical contractor about the location of the new VAV boxes. Although a drawing was issued to the most relevant person, there was little thought given to the other subcontractors who might have a smaller stake. In this case the need to modify electrical runs, lighting, or wiring locations for these new box locations is needed by both team members. The process of sharing single response documentation with particular team members usually is a recipe for disaster and often avoids the ideals of increased collaboration by focusing on single tasks and not the project as a whole. Unfortunately, this method of information sharing has been the cause of change orders and unnecessary costs that are passed on to the owner from lack of team coordination.

IMAGE COURTESY OF BRAD HARDIN

Figure 5.35 Job site trailers as the main hub for communication during construction

The Virtual Job Trailer

So, how does BIM work in a job trailer, and what new processes and technologies make the most sense and are most effective for the project? To start, I will talk about the job trailer as a conference room (Figure 5.36). The primary difference between CAD and 2D projects and BIM is that the information is contained on computers. That said, a job trailer for a BIM project will need to have the ability to display the model, sheets, and views generated from the model in the trailer. Many times this is solved by including a flat-panel screen in the trailer. The cost of a flat-panel television can range from $500 to however much you want to spend. These are usually a very affordable option and can be reused on future projects, so they're worth the investment. However, if a flat-panel display is not an option, a less expensive option is a larger computer monitor or projector. These can be purchased anywhere from $300 to $3,000 dollars and are just as effective as an LCD or plasma display; they might lack versatility for some higher-definition or display-size functionality. Other equipment that a BIM-ready job trailer might need to have is a wireless router that links with the flat-panel or monitor for ease of use when multiple laptops are being used in a meeting. This eliminates the need to swap out cords and is a great way to effectively communicate using the computer. This method is more environmentally conscious as well, and comments generated during the meeting on a model may be combined into the same tool containing field comments for future direction and resolution, without printing.

IMAGE COURTESY OF BRAD HARDIN

Figure 5.36 Field coordination is where all the physical pieces come together.

The trailer's second function as a plans and specifications hub will continue to evolve with BIM; not only will paper plans and specifications be replaced, but trailers are now being outfitted with servers linked to wireless networks for their field personnel to utilize tablet PCs in the field. The term *tablet PCs* is used loosely here, and although they are becoming more prevalent, many of these tasks can be accomplished on a laptop computer with wireless Internet access. The keys here are mobility and Internet access. According to George Elvin's book *Integrated Practice in Architecture*, the use of wearable PCs in the field can reduce rework on a project by 66 percent when compared with more traditional methods. As rework currently accounts for about 12 percent of total construction, tablet PCs could reduce costs by as much as 8 percent overall. The key element to this success is the remote access to network information, which allows field personnel to access the most current data anywhere on the job site. This is particularly effective on larger projects such as stadiums, warehouses, casinos, hotels, and larger civic projects where even walking back and forth to the job trailer is an ineffective means of project coordination.

In addition, using tablet PCs for a project enables the subcontractors performing the work to scan and locate building equipment. Software companies (Vela Systems) have developed software, which enables the use of RFID (Radio Frequency Identification Device) tags. These tags act similarly to a barcode and can be scanned from a scanner pen attached to a tablet PC in the field to pull up the information for that building component. This further assures the accuracy of construction the first time in order to limit costly errors later.

Tablet PCs can also be enabled without a job site server, through wireless Internet access, where users can tap into their own network to access data stored on the network back in the main office through application virtualization. Through the use of tablet PCs, an Internet connection, and viewing software, you can limit trips to the job trailer, and the efficiencies of sending and receiving live project data is much more streamlined.

Lastly, the ability to automate project updates is key to construction administration. Using RFI software mentioned earlier, the model may be distributed in real time from a single location, which is invaluable to a team. Many times the most valuable part of BIM is the *I*, or the information, but the fact that it is housed within a three-dimensional representation of this information in a single file makes access much easier. The goals of BIM during field administration is to keep the amount of files to a minimum, limit potential errors and miscommunication on a project and centrally locate all associated files. For example, if a project superintendent needs to generate RFIs and the project is using Acrobat for field coordination, then the RFI should be generated in PDF and merged into a file for all to review. As discussed, this process has limitations and is not currently possible in all cases.

In essence, a construction manager wants their team looking at the same up-to-date information at the same time. This frankly is not possible in a CAD process for a number of reasons. The first is the number of CAD files that would have to be used requires significant coordination resources. Another is the information in static sheets is usable but, well, static. This is unlike the BIM files, which can stay current and reflect model changes in all of the sheets throughout the project. BIM gives a unique opportunity to continue to house the information and data flow for a project in a single location, with work being accomplished using one or two other tools instead of numerous files sent different ways and stored at different locations. Using a BIM process, the virtual plans and specifications hub becomes a much more integral part of the equation when compared to old processes. Yet unlike old processes, there is an investment of time and money required to make the new processes work. The key to utilizing this technology is using it where it makes sense.

The last process of BIM in regard to onsite construction is the function of the job trailer as a communication hub. Face-to-face interaction and collaboration are often the best mechanisms for project results, but, some of the other tools, such as the display panel in the trailer, might provide the team with an effective way of videoconferencing with architects, engineers, owners, code administrators, consultants, and so on. Just as tablet PCs can effectively provide a means of updating the team on progress meetings via email, photos, or live calendars, videoconferencing can keep everyone linked in. Basically, a communication hub is still most effective through direct human interaction and collaboration, but BIM offers the possibility for enhanced virtualization.

The advantage to having a job trailer enabled with different BIM tools that have access to accurate BIM data is that all users are referencing the same set of information. This doesn't solve all the problems of a CAD-based process, but it does begin to streamline them, because fragmented and dislocated data is replaced by centrally located data that is accessible by everyone. This type of solution is truly unique to a BIM process and gives all stakeholders access to all the information relevant to a project, whether it be the central model in its native software, the compiled RFI file or the analysis model. By creating this digital "hub," stakeholders are able to stay current on their project miles away or, as is often becoming the case, halfway around the world. This type of collaboration with the fundamental understanding that one or two sources for all of the information in a project are correct and up-to-date gives the team confidence and promotes further collaboration. As teams weigh in on changes that might affect them, they are also confident in the quality of information to seek coordination with other trades.

BIM and Site Coordination

Building information modeling and site coordination can involve sustainable site management, building component tracking, commissioning, GPS material location, scheduling, and so on. Because this is a broad field and there are many tools for site coordination, in this section I will analyze how BIM can be utilized to assist in site coordination specifically. Site coordination is unique because in many cases using BIM technologies goes further into more specific technologies that assist in a more coordinated project and integration. In the future, as BIM continues to evolve, there will be more tools that dovetail into the model and documents that make construction management run smoothly.

IMAGE COURTESY OF BRAD HARDIN

IMAGE COURTESY OF BRAD HARDIN

Site coordination refers to the organization of the site, materials, equipment, safety, and site security. Earlier in this book I outlined how to create a site logistics plan. Now I will show how to put it to use in the field. You can use this plan for many purposes, from safety controllers such as OSHA to governing authorities to subcontractors to material providers. The site logistics plan will play a vital role in creating better communication and a safer project; the plan is often housed in the job trailer and posted on the wall for reference. This assures that all team members have a visual understanding of where material lay-down areas, site accessibility, parking, and building access are. Although the site logistics plan is a static image of the site, there can be variations of the plan because of sequencing, scheduling and different phases of the project. For this reason, it is not uncommon to have more than one logistics plan throughout the course of a project.

Using the site logistics plan is especially critical when the material being delivered on a job site is large or numerous. Combined with a tight site, such as in an urban setting, material lay-down coordination can be a full-time job depending on the size of the project. A BIM solution to this complex site coordination could be a sequencing animation or a series of site logistics plans. The question of what the material is and where the material is going can become a concern on-site as materials begin to pile up.

Current technologies have begun to leverage the use of RFID tags that can be placed on building components and scanned using a handheld computer, which brings up the building information about the component and where it is to be placed. This is particularly effective when dealing with very complex structures or buildings that require multiple phases of construction. RFID tags can also be enabled with GPS locators. Paired with RFID-enabled software, the project manager can view where the building components are at all times. This technology paired with a BIM model provides material component information through the use of handheld computers and scanners as well as the location of components through GPS.

> *Individual objects scheduled for arrival on the construction site are tagged at the fabricators using bar codes or radio-frequency transponders (RFID). The encoded information is scanned directly into a portable computer and wirelessly relayed to a remote project database. A database query returns graphical representations (ex. computer aided design [CAD] information, or virtual reality mark-up language [VRML]), models of scanned objects and additional information as appropriate. These models, coupled with user-friendly web browsing software, guide field workers through the acquisition of key fiducial points using scanning devices integrated with GPS technology to determine an object's position and orientation. Pallets of materials can be more easily tracked; even permanent equipment like HVAC systems, pumps, motors, etc.*

can be tracked from their arrival on the site to installation. RFID tags attached to equipment could contain the complete maintenance history of the tool or device.

The Smart Construction Site of the Future, www.technovelgy.com

This technology may also be used for construction equipment, cranes, bulldozers and lifts, as well. Virtually anything that a tag can be adhered to is able to be scheduled and located. As a result, using RFID and GPS tags on equipment is particularly interesting to construction managers who own and operate their own equipment. The ability to place an RFID tag on a piece of equipment can report information about the equipment including oil changes, routine maintenance, and issues to a field laptop instantly, which could potentially make the use of RFID tags more widespread in the construction industry. In the future, RFID tags could very well be tied to a web calendar that would send maintenance personnel reminders of equipment that needs to be serviced, and in turn, the maintenance personnel could identify where the piece of equipment is to do the work on it.

Site safety and security are very important issues when dealing with any construction project. Although the use of BIM is important when there are potentially hazardous areas cited in the logistics plan, there is also technology, such as web-enabled security cameras, that go beyond the realm of BIM. Some construction projects provide the use of security cameras on-site to not only be provided as a safety measure but to analyze the progress of the structure for project team members to view over the Web. Currently, there are a large number of security camera companies in the industry, yet the functionality of being able to move and position the camera remotely allows for functionality that didn't exist in earlier decades. This additional degree of visualization is a way for other stakeholders to gauge progress, potentially foresee any constructability and safety issues, and address them using this tool over the Internet. Some companies find it interesting to overlay both the BIM and the current construction snapshot views over each other and compare the actual to virtual level of construction completion as a way to see how and if the project is staying on track.

Fabrication

The use of BIM in fabrication has changed the way many parties in the construction industry operate, and it continues to have the potential to change the industry holistically. Since a BIM file contains parametric modeling information and many fabricators use 3D models to build their components, there is a great opportunity for information exchange between the two. These are some examples where 3D fabrication exists:

- Structural steel: Columns, beams, rebar, joists, and bracing
- MEP components: Ductwork, conduit, equipment, piping
- Precast concrete: Custom shapes, patterns, and reinforcing

- CIP concrete: Reinforcing layouts, component numbering
- Specialty items: Custom handrails, brackets, and sunshades
- Glazing systems: Customized assembly systems, connections, details
- Other items: Furniture, signage, site features, sculpture

As it pertains to a construction manager, the ability to receive shop drawings from BIM-enabled subcontractors during the documentation stage and compile and test that information in a construction model is an incredible resource. Although the field of subcontractors capable of producing fabrication models continues to grow, some haven't embraced this technology yet. In reality, subcontractors have the most to gain in a BIM process through the automation of production using Computer Numerically Controlled (CNC) machines, which reduces upfront coordination time and equates to a reduction in rework overall. Additionally, the fields that offer BIM-capable applications continue to grow in areas that require off-site fabrication. In *The BIM Handbook*, the authors explain the difference between made-to-stock (drywall, fixtures, studs), made-to-order (windows, doors, hardware), and engineered-to-order components (structural, MEP, custom concrete, and specialty items). Overall, BIM is most applicable to the engineered-to-order (ETO) components of the three categories.

The ETO category includes items such as structural, mechanical, and specialty items that are unique in design and construction to a particular project. Although they may be built using standardized parts and connections, the actual layout and design of these components are unique. Engineered-to-order items must be designed, sized, tested, analyzed, fabricated, and installed to exacting standards. The coordination of these components is often the most critical element of a project because the components deal with a number of relative unknowns through custom fabrication and because they can affect multiple systems. For example, a custom-angled curtain wall system will have many more unknowns than a standard rectilinear storefront system. Components such as customized supporting structure, custom flashing and waterproofing, and custom fabricated mullions all are systems that require unique solutions to be resolved. The proficient utilization of BIM coordination during fabrication carries into construction. With this in mind, it is best to analyze the BIM files prior to construction to reduce the issues in the field. Much of this can be achieved by completing a clash detection analysis with other systems or a sequencing clash (to verify that the size of the preassembled components can be installed) by using the fabricators file. Because the fabricator's model is a true representation of what is going to be sent to the machines to construct, it is the most accurate model a team can utilize. This BIM-as-built approach to documentation is a logical progression of the VDC BIM. As you increase the information and integrate more detailed models, you increase the accuracy of the project and reduce overall unknowns.

The concept of BIM-as-built directly applies to lean construction coordination prior to execution. Lean production and procurement are parallel efforts because BIM fabrication is enabling the following to take place:

- Offers the ability for the construction team to procure and fabricate components accurately (earlier if need be)

- Limits the amount of rework and hours in-field

- Less paper waste because of digital submittals and coordination

- Improved team buy-in and trade coordination

- Less overage on ordering and waste reduction

- Reduction in total costs for engineering and shop drawing time

Obviously, using BIM for fabrication lends itself to an integrated process because it simply wouldn't be possible without the early involvement of the subcontracting/fabrication team. Although different opinions about the direction of model coordination exist, the overall thinking seems to be that the process of refining the VDC BIM to a fabrication level of detail is most effective (Figure 5.37) prior to commencing construction.

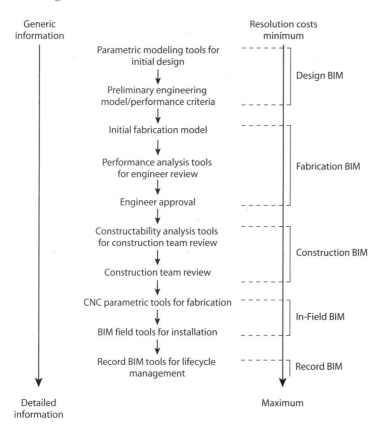

Figure 5.37 Refinement of a BIM from design to fabrication

Although some of these processes might overlap, such as a constructability and engineering reviews, the overall refinement of the VDC BIM to a fabrication model is developed in tandem with the actual project design. For example, the tools that an engineer might use to test a shop model such as Tekla or SDS2 for structural and IES or eQuest for mechanical are tools specific to that team member. And it's important that an engineer be focused on verifying the performance of the systems and use the tools available to complete this work. Other analysis such as constructability issues where BIM can be leveraged for clash detection, clash sequencing, increased schedule visualization, clearance clash detection, reverse clash detection (deviation testing), and other trade coordination can occur in tandem with the engineering analysis though on separate software. Both types of testing and analysis are completed and overlap for time's sake; however each have different results sent to the project team. This evolution of the model's level of detail and the refinement from analysis lead up to the construction phase of a project. In the field, the resulting product should be a useable accurate tool, which can then be used to verify the accuracy and installation of the entire project.

The level of detail a subcontractor and fabrication team are capable of allows the construction manager to associate even more advanced 4D scheduling capabilities to the models to further refine and coordinate the construction. Because the building has been constructed virtually, the ability to produce more detailed scheduling coordination is possible as well.

Navisworks Sequencing Clash Tutorial

Navisworks not only allows for users to create clash detection reports but also allows for them to create clash detection reports tied to a "simulation." These reports are extremely valuable when different construction is going to take place concurrently, such as a medical room receiving a large piece of equipment such as an MRI machine in which the room needs to be built around the piece of equipment. The ability to show three concrete walls and the roof going up, the piece of equipment being hoisted in, and then the last wall being erected is extremely valuable to a construction management team. But what about the ability to show the below-slab coordination that might require a slab to be delayed being poured or might require a housekeeping pad to be poured after the main slab of the equipment room so the machine is mounted at the correct height? Although construction scheduling animations tell the story as an overview, the sequencing clash allows the construction of such a facility go more smoothly.

In this tutorial, you will create a clash report that informs you that you are beginning the concrete pour for your housekeeping pad prior to the time you need for the main slab to complete curing in the basement of your facility.

Creating a Sequenced Clash Detection Report

1. Open the `construction-sequence.nwf` file in Navisworks Manage from (www.wiley.com/go/bimandconstruction).

> **Note:** Notice in this file there are two linked models: `arch-model` and `housekeeping pad`. This exercise simulates two concrete pours in the basement of the proposed building. So, in this example, you have a slab on a grade that needs to be poured, and you need to verify that the curing time, which in this example is 30 days, doesn't overlap with the timing of your new pour for the housekeeping pad.

2. Open the Clash Detective window.

3. In the left window, select the 6″ Concrete Basement Floor item that appears under `arch-model.nwd` > Basement.

4. Select `housekeeping pad.nwc` on the right (Figure 5.38).

5. Change the clash type to Hard, and change the tolerance to 0ft 0.00.

6. Change the Link setting from None to TimeLiner, and leave the Step (Sec) setting at its default.

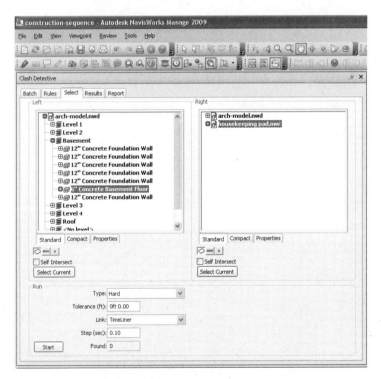

Figure 5.38 Establishing the model clash parameters

Note: In this example, you're using the Hard clash type because you are making sure that any model components abutting or embedded in the slab floor do not intersect. This type of clash may also be done using the Tolerance tool, with a user-defined tolerance setting to verify clearance between two objects as they are constructed.

7. Click Start.

8. On the Results tab, you can view the clash and highlight the clash. (This example should have found one clash.)

You will see that you have a conflict in pour times between the slab floor and the housekeeping slab (Figure 5.39). Additionally, you can see that the date of the clash appears in the upper-left corner of the model view window and that the TimeLiner window's Simulate tab has already advanced to where the clash is occurring (Figure 5.40).

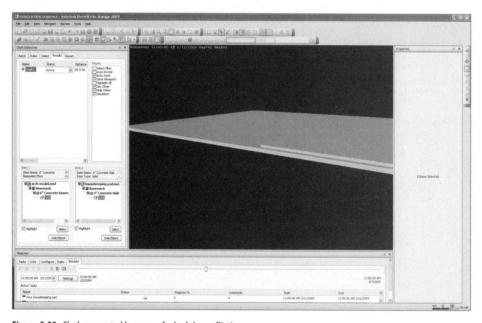

Figure 5.39 Clash generated because of schedule confliction

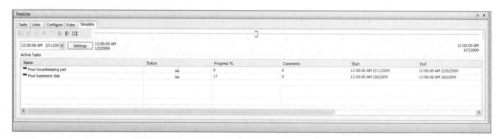

Figure 5.40 The date the sequencing clash was found is displayed.

Although this is a simple example of how a sequencing clash detection works, it shows the value in being able to verify for complex structures that critical path construction elements aren't interfering with one another. As construction progresses, it is also a means of viewing which components are to be installed in what order. Lastly, the sequencing clash can be used with mass model elements that indicate material laydown areas, which can then be tied to procurement schedules to verify successful site coordination. The sequencing tool is powerful and, when used in tandem with a solid BIM strategy, can be used to better coordinate and create efficiencies in the amount of time it takes to review 4D construction issues.

BIM and Punch Lists

Punch lists are an especially interesting area in regard to BIM, and tools are currently available that could streamline processes and embed punch list information within the BIM model. With the use of handheld computers, construction management personnel are now able to walk from room to room and provide detailed information about the punch list items. These items are then tracked and logged, much like an RFI. These are some of the options currently available:

- Creating custom room schedules in the modeling software (Revit, ArchiCAD, and Catia)
- Using software such as Acrobat to create specific markups within sheet *and* model files
- Using Navisworks or similar software to automate numbering and directly host comments to parametric objects for review
- Using plug-in or export software that allows users to see an exported or linked version of the BIM to be tied to external software that hosts comments

To understand the value of a BIM-driven punch list, you must understand how current processes and technologies exist. In a typical process, punch lists are generated in different ways. The first means is by notations on construction documents, which the architect or contractor has printed. These documents might or might not be the most recent accurate drawing data but are used with the intention that they will be used as reference only as the architect and owner highlights issues that need to be fixed prior to substantial completion. After the architect has generated the handwritten list of issues, the architect either scans and emails or gives a copy to the general contractor who then starts fixing the issues.

Another means of generating punch lists is creating a room numbered spreadsheet either in Excel or in another format that assists in getting issues resolved by the general contractor. The general contractor then makes a number of copies and distributes them to the subcontractors at progress meetings. When the subcontractor has fixed an issue, the general contractor marks the item as completed and hands the completed list over to the general contractor. When the architect completes the second walkthrough, the architect lists the items as completed or not acceptable. The general contractor then notes this in a spreadsheet. However, as the architect is walking the site the second time, new issues might arise. Or if the project is a 30-story high-rise and the punch list is to be accomplished by going from floor to floor, then the amount of data may become overwhelming, especially if the means of documenting the punch list items was through handwritten sketches and notes.

In its simplest form, BIM is a database. This database of information is represented by three-dimensional objects yet at its root is still a database. That said, you can link the database of room information to punch list items in a number of ways. The first is using the native modeling software to create customized room schedules that associate comments and tags with the current BIM file. For instance, Figure 5.41 shows an example of a simple customized room schedule created in Revit that could be utilized by field personnel in tandem with the relevant floor plan. Ultimately, using the native BIM then becomes residual information within the record BIM model upon project delivery. This has a couple of advantages. The first is that the owner or owner's representative now has the ability to use, for example, a centrally hosted Revit file and add comments and issues in tandem with issues generated from the architect. Other project members in this process are able to contribute to the punch list as well. More important, multiple users with the native software can update and review the status of items from a centrally hosted file.

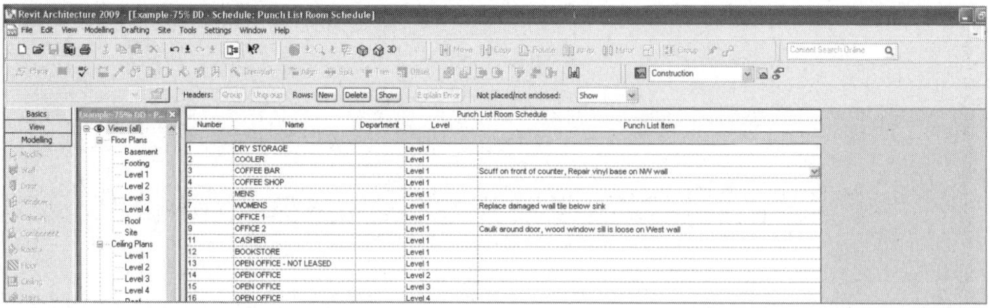

Figure 5.41 A simple example Revit punch list schedule

Another advantage to this type of documentation is that the facility manager now has an issue history of the facility. This can be useful because the punch list now becomes aggregate information within the record BIM for reference. Although this may not be of particular importance to some, the ability for it to be referred to in this type of punch list strategy is very important to a facility manager tasked with maintaining the structure's information history.

A disadvantage to this type of strategy is that all users wanting to plug in to the central file often need to have a copy of the native modeling software. Additionally, there can be a high degree of customization to creating a punch list room schedule.

Although the example in Figure 5.41 shows a base-level spreadsheet, a punch list would rarely be this simple and would take some time to create.

Another option for linking the database of room information to a punch list is to use Acrobat to address, log, and track punch list items. In Acrobat, you could use the composite model and sheets and begin to tie issues to rooms within the BIM, using the notes or commenting tools already available. The advantage of creating the punch list through Acrobat is that you can link the punch list items to this (ideally) singular file. Another big advantage to this method is that all the information is still contained in a single file and can be accessed by all the team members via the Adobe Review Tracker tool during construction. Using Acrobat requires that users have a base copy of the software to create comments, but the cost of this software is often far less than the cost of modeling software so is a cost-effective means of completing a punch list (Figure 5.42).

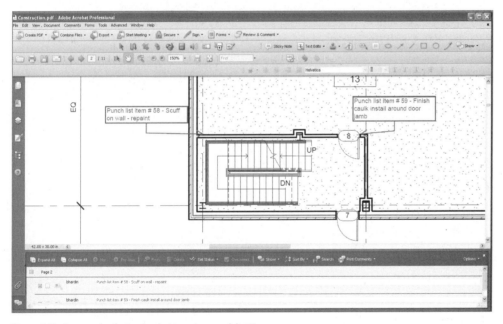

Figure 5.42 An example of using Acrobat to review punch list items

A disadvantage to using an Acrobat file is that it is unlinked information. Although the construction manager can use the notes or commenting tools for issues that are addressed during walkthroughs and room inspections, that information cannot be fed back into the native modeling software. Another disadvantage is that this method requires some schedule customization to be effective, though once the schedule is built, it can be copied and used for use in future projects.

Finally, you can also use Navisworks to directly mark and comment on the BIM model (Figure 5.43). This method of coordinating punch list items is very effective and relies on the software's ability to automate the creation of sequential comments, view specific commenting, and the review tool within Navisworks to complete the project to closeout. Navisworks is particularly effective if already being utilized for in-field RFIs, because the team will have a base understanding of the software and will be able to use the familiar interface to sort through the final issues of a project. Navisworks doesn't necessarily require any customization, and although external databases can be inserted into Navisworks, the default tools are straightforward enough to be effective.

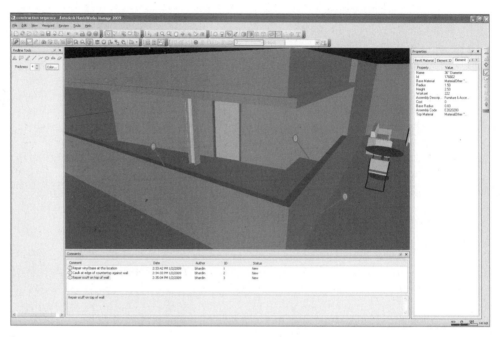

Figure 5.43 Example of Navisworks being used for punch list coordination

A disadvantage to using Navisworks is that in order to be commented on, an object must be modeled. There are ways around this such as selecting a component behind where the model would be. For example, you could select a spot on a wall opening where the windowsill is and create a comment there. Another disadvantage is that the team needs licenses; although these do not need to be as robust as the full-blown Navisworks Manage tool to be utilized, there are still additional costs. However, as BIM continues to develop toward a BIM-as-built scenario, Navisworks seems to be a worthwhile investment for a firm doing multiple projects in BIM.

Current BIM tools are effective; however, punch list software will continue to develop that provides users with more robust tools for managing information. Recommendations for future iterations of punch list software would be to create editable databases that remain linked to the model. Because everyone on a team will not necessarily be familiar with the ins and outs of all BIM creation software, allowing users to easily create and update punch list items would be very effective. Although these tools have room to grow in sophistication, current tools are a prime example of how effective the BIM database even at a basic level can be, depending on how you want to use it.

Conclusion

This chapter has covered building information modeling and clarifications, how to train the superintendent, set up a BIM enabled job trailer, field construction, and punch lists. All of these topics are specific to BIM processes; however they are not by any means the only coordination that can be accomplished on a job site. In a way, this chapter covers the basic "how to" of BIM during construction administration and it shows where there are current solutions and shortcomings alike. Conceptually, what is important is that the evolution of BIM from design to fabrication, to construction should continue to become a more refined and accurate tool. The model should only be replaced by a more accurate and detailed model.

A BIM process during construction administration is continually being defined as well. Every day new case studies and methods rise to the surface and shape what this exciting technology can hold in store. While the potential is great, it is important that tools are developed in the area of construction administration that address more accurate and linked modeling and analysis strategies.

The RFIs have been answered, the punch list has been completed, the owner is getting ready to take possession of the space...so now what? The value of BIM technology will continue to be in its ability to house important life-cycle data that offers information to the all stake-holders. For this reason, creating an accurate VDC BIM during construction aids in the accuracy of other analysis tools such as those discussed in the next chapter.

BIM and Sustainability

In his book The Triple Bottom Line *(Jossey-Bass, 2006), Andrew Savitz shows how current and future companies will become profitable based on the effective management of the triple bottom line, which includes economic, social, and environmental issues. The chapter shows how profit is tied directly to the way we respond to properly managing all three. In this current economy, many are looking toward solutions to create jobs, create profit, and "lean" their teams. Sustainability has become a large part of this response. This chapter reviews the current status of the industry in regard to its drive toward a more sustainable means of constructing as well as future opportunities and developments that create a positive synergy between technology and the environment. In this chapter, you'll learn about the following topics:*

In this chapter

Overview of BIM and sustainability

USGBC and LEED

Green BIM and preconstruction

Sustainable construction

Salvaging and recycling

Overview of BIM and Sustainability

The face of design and construction in the past decade has shifted the way we are using natural materials and manufacturing them into building products to construct buildings. Our use of the world's natural resources such as water, metals, and wood has resulted in an increase in material costs in recent years. For example, from 2007 to 2008, steel prices have almost doubled (www.steelonthenet.com/prices.html). And as the world's population continues to grow at a rate of about 203,000 people per day (http://en.wikipedia.org/wiki/World_population), the question has been asked, "How are we going to build more sustainably, creating better buildings and using a better means of construction?" One of the main answers continues to be technology. Whether it is BIM, GPS, GIS, nanotech, green tech, or alternative technology, the industry looks to the tools in the technology sector to provide access to data, conditions, and answers. In this century alone, we have been provided with access to more information than any previous century. As such, the real challenge is now figuring out how to manage and use that data to formulate and make better decisions about the environment. Green building standards have seen a significant rise in popularity, evidenced by the United States Green Building Council (USGBC) and the Living Building Challenge (LBC – Cascadia Green building Council). In fact, according to the McGraw-Hill Construction Analytics' SmartMarket Trends Report 2008, "The value of green building construction is projected to increase to $60 billion by 2010." So, what tools are architects and designers using to create greener buildings through BIM technology, and moreover, what tools are available to contractors to construct and manage in a more eco-friendly manner?

Currently, architects and engineers are using BIM to help accomplish a variety of tasks more efficiently than in the past and some that might not have even been possible. The book *Green BIM* by Eddy Krygiel and Brad Nies is dedicated specifically to BIM, green construction, and design strategies and is an excellent resource to dig more in depth into how project teams are utilizing the power of BIM to create a truly informative tool throughout the design process. As an overview to the book, the way a team now designs and interacts with other team members is now different because of an increase in the amount of information and because of an increase in efficiency in technologies to use and share this information. For example, architects can input data into the BIM file that geographically locates the building and digs up information that helps the design team understand issues related to climate, place, surrounding systems, and resources. Designers can then edit and reorient a building on a site using real coordinates to reduce the resource needs and use more efficient solar orientation. During schematic design, BIM provides an architect with the ability to analyze a building's mass and form to optimize envelope and balance glazing ratios. Engineers also use BIM files to reduce energy demands through energy modeling, which uses the 3D model to calculate light reflectance and penetration. Contractors can analyze site conditions, including wetlands and protected habitats, and use the site model to coordinate logistics better to eliminate potential issues. Additionally, in the future when there is an existing model of a facility, a contractor will be able to quickly quantify what materials are coming out of the building to facilitate reuse or recycling efforts. Subcontractors can utilize BIM technology to reduce waste and combine shipments to further reduce carbon footprints. Ultimately, the BIM file provides an array of resources needed early to inform the project team about many of the sustainable issues pertaining to the project. So, how can you use BIM to better the way you design and build structures?

USGBC and LEED

To begin, the strategies of a BIM process directly correlate into that of a sustainable design strategy. Strategies such as early goal setting, team coordination, responsibility matrices, construction efficiency coordination, and technology to provide better answers are included in both. So, how do you integrate a sustainable modeling strategy into the proposed development of a BIM process? To start, we need more integration between the contracting team and the design team; more specifically, identifying what purposes the contractor can use the BIM file for. These may include a number of tools throughout the course of a sustainable construction project that are analyzed in tandem to model development. Prior to beginning any of these new processes, though, contractors need to become educated about green building practices. One such way to begin learning about eco-friendly construction is to study the information standards compiled by the USGBC and in particular the Leadership in Energy and Environmental Design (LEED) reference book. The LEED manual is a series of credits that provide scalable and real goals for new construction projects to attain. For LEED certification to be attained, the project team must achieve credits from the following categories:

- Sustainable sites
- Water efficiency
- Energy and atmosphere
- Materials and resources
- Indoor environmental quality
- Innovation and design process

Since it was founded in 1993, the USGBC has seen exponential growth and now consists of more than 18,000 member organizations throughout the United States, as of December 2008 (Figure 6.1). Although this speaks as a testament to the importance that building professionals see in green building and design, this increase is also being brought about by owners who find it's easier to sell a "green" condominium to an educated buyer, who realizes energy savings in an energy-efficient building, and see many of the other benefits as well. Additional savings, such as a reduction in water use, are good economic sense and will continue to be a driving factor for future generations to build upon a standard of designing and constructing sustainably.

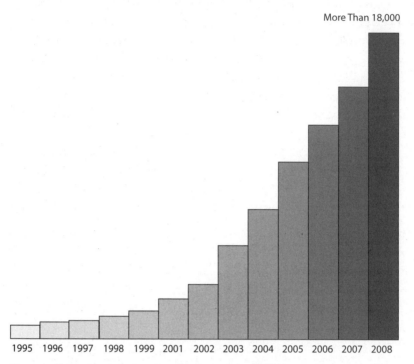

More Than 18,000

1995 1996 1997 1998 1999 2001 2002 2003 2004 2005 2006 2007 2008

Figure 6.1 The number of USGBC members has grown significantly in the past decade.

Contractors have begun to find that not only having BIM services in-house but also gaining LEED accreditation means they are better qualified when competing against those who aren't accredited. In fact, many contractors have begun creating a market focused on green building construction and have started to specialize within the industry to further delineate themselves from their competition. LEED accreditation is open to anyone who desires to design and build sustainably; the tests are available to owners, architects, engineers, facility managers, consultants, contractors, and individuals. Although LEED first started as a grassroots effort within the architectural and design community, it has now spread to the owner and construction sectors. Many industry analysts believe that green construction and design methodologies will no longer be a fad but will become a commonplace means of construction. Essentially, green construction will just become good construction practice.

Building information modeling as it relates to LEED projects can be used a number of ways. For instance, the construction manager may use the model to assist in more accurate procurement of materials for the project, provided the model has been created to a high level of accuracy. More accurate material ordering means less waste on a project and less cost for materials not used on the job. Another resource for sustainable BIM use is the ability to analyze performance of a virtual construction and then test the actual performance after construction. This is particularly valuable for the AEC team that must verify a building is performing as it was designed. For example, models can be virtually tested for heating and cooling loads and airflow using computational fluid dynamics (CFD) software. The model becomes the performance benchmark, and during commissioning after construction, when the physical building's performance is analyzed, the team can measure actual performance to the estimated performance. Building information modeling also allows for a number of calculations and simulations, such as animating a sun study for a year to verify shading device efficiencies, rain water runoff, and site analysis. The list goes on; however, as LEED measures success and certification based on credits, BIM allows a project team to virtually construct a model and make better assumptions as to the number of credits the project is eligible for. Another added benefit is the flexibility BIM offers a design team who previously had to complete calculations with each different design iteration. Using BIM technology, users can now update information quicker than before; when used with analysis software, many project teams are pushing the boundaries of possibilities because of the speed with which they are able to analyze multiple options to arrive at the best solution.

In addition, many code enforcement agencies are looking at their own standards and implementing green standards into their code manuals for the industry. Such organizations as the ICC have adopted many green practices and have implemented them into their codes. Recently, the ICC created the Sustainable Building Technology Committee (www.cela1.org/pdf/ICC_eNews_073108.pdf) to continue to find ways through code enforcement to standardize green initiatives in the construction industry. Many organizations have shifted to promote and integrate green building standards, but there are still those who remain unresponsive to the shift. Unfortunately, those groups and companies will be left behind while their competition increases their green building portfolios and familiarizes themselves with this movement. In the end, every construction company will be required to adapt their organization to comply with any industry green standards being adopted into building codes and local ordinances around the world.

Green BIM and Preconstruction

So, how can you use BIM during the preconstruction phase? To answer that question, let's focus on three main areas directly connected to BIM and sustainable construction: material selection and use, site selection and management, and systems analysis. These all are heavily weighted and are critical to achieving LEED certification for a project. In addition, all three can use the model to be scheduled and tested. The effort to construct a sustainable building facilitates a change in process. As before, the value of a more integrated process is in what everyone brings to the table, such as industry experience to assist the design team in making sustainable decisions, questions about project goals, and input into the process from a real-world construction perspective. Team members and consultants can offer a wealth of other information such as available material resources within the required project radius, means of constructing alternatively, and utilizing regional resources to more efficiently build the project.

Material selection is one of the most important factors when designing and building green projects. Buildings use 40 percent of raw materials globally, which is 3 billion tons annually (according to Worldwatch Institute's "Worldwatch Paper 124: A Building Revolution: How Ecology and Health Concerns Are Transforming Construction" report from 1995). Although that number seems staggering, what's more unbelievable is that the amount of construction in the world is anticipated to not slow down but to grow at exponential rates. This shows the importance of defining how we will use the earth's natural resources most effectively. Although using technology will get us only so far, it will take a concerted effort from everyone on the planet to use what we have responsibly. Let's face it—the world isn't making any more stuff, and we need to realize that all of our actions have results.

How do you begin selecting the right type of materials? One of the answers is research. In the past, construction companies might have just been concerned with square footage and volume of material with no thought given to the way the material was processed, where it was shipped from, or what chemicals it contains. The fact that materials might be emitting harmful toxins into the air after installed or that paint might contain chemical compounds that could be harmful to children escaped the watchful eyes of many designers. Times have changed. Now, information in the form of online resources, print materials, and research experience from industry professionals and third parties has begun to shape the way materials are selected in the AEC industry. Contractors need to think about where materials are coming from in order to meet material location radius requirements. The construction industry also needs to think about fuel consumption during shipping, material storage, and dust in a whole new way. A whole or more *holistic* way of thinking is fundamental to contractors who understand green construction. Contractors also have a range of tools to use for finding, inputting, and extracting information from the BIM file. Additionally, contractors who are using BIM can build a historical database of estimated cost, performance, and resulting products from the BIM file with actual results. This information can be extremely valuable as green design and construction practices are refined and sustainable case studies leveraging BIM continue to increase.

You can use BIM during the material selection stage to add information to building components; for example, inputting percentages of recycled quantity information, such as the ratio of fly ash into a concrete mix or recycled rubber material in tile carpeting. Other relevant information might be reclaimed material quantities and locations for separate scheduling or inputting manufacturing location information to gauge what materials might have been regionally sourced. Since the BIM file updates material quantities and volumes as the building is modeled, you gain a very accurate understanding of what percentage of materials are contained in the latest version of the model. This allows you to quickly analyze whether you are achieving the LEED credits or project goals you have set out to attain on the project. In addition, by using material quantity schedules within Revit, you can create separate schedules so you can input material information into the model. Later in the project, these material percentages will ease the process of LEED reporting and takeoff to generate comparisons of both cost and material weight to other materials.

Custom Schedule Tutorial

For this tutorial, you will create a custom floor schedule in which you will input a custom field to indicate the amount of recycled content percentages of your building and derive a total volume of recycled content of your floors. To begin, open Revit, and open the 100% CD file.

Creating a Custom Schedule

1. Click View > New Schedule/Quantities (Figure 6.2).

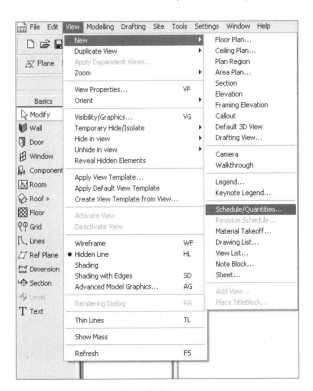

Figure 6.2 Creating a new floor schedule

2. In the category window, select Floors.

3. On the Fields tab, add Family and Type, Area, Level, Volume, Perimeter, and Comments.

4. Click OK to create the schedule (Figure 6.3).

   Now that your schedule is created, you need to add a new parameter and add a calculated value to that parameter so that you can show the recycled content in the concrete floors.

5. Right-click your schedule, and select View Properties (see Figure 6.4).

6. In the Other field, select the Fields category again.

7. Now select Add Parameter.

8. Under Parameter Properties, change Parameter Type from Project Parameter to Shared Parameter.

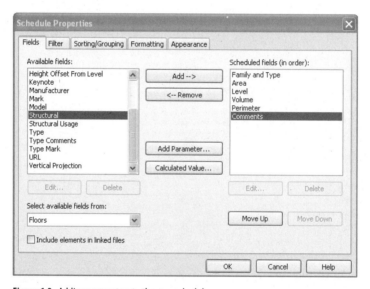

Figure 6.3 Adding parameters to the new schedule

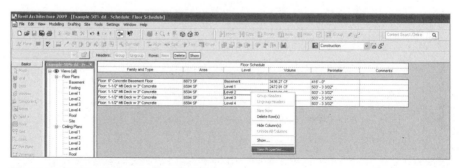

Figure 6.4 Editing the view properties of the schedule

9. Click the Group Parameter Under drop-down menu, and select Materials and Finishes (Figure 6.5).

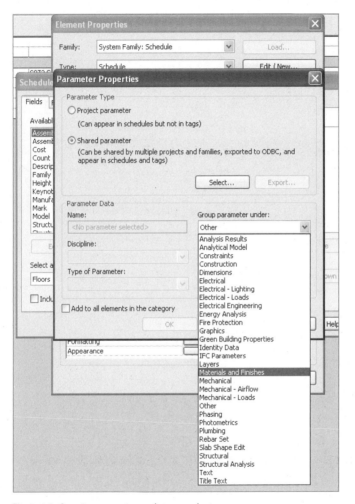

Figure 6.5 Grouping parameters under a named category

10. Now click Select.

11. In this window, click Edit.

12. Click Create at the top right of the window.

13. Under File Name, enter **Recycled Material Parameter** (Figure 6.6).

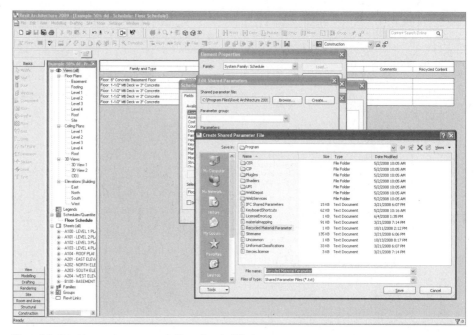

Figure 6.6 Attaching a custom parameter file to the schedule

14. Click Save.

15. Now under the Groups category, select New.

16. Type **Recycled Content** (Figure 6.7).

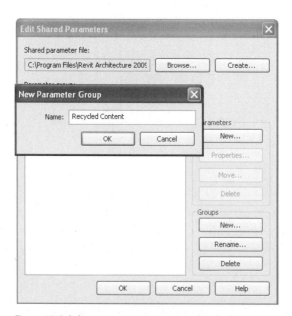

Figure 6.7 Labeling a new parameter group to the schedule

17. Click OK.

18. In the Parameter Group drop-down menu, select Recycled Content.

19. Now Click New under Parameters.

20. Under Parameter Properties, fill in the name as Recycled Content. Under Discipline, enter **Common**, and under Type of Parameter, select Number (Figure 6.8).

21. Click OK until you are at the Schedule Properties window.

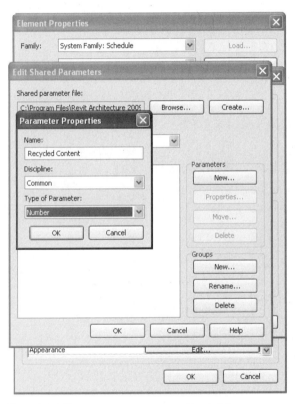

Figure 6.8 Adding a new parameter group to the schedule

Now under the scheduled fields, you can see the Recycled Content category you just created. Although this seems like a significant undertaking, you made the Recycled Content category a shared parameter, so you will have to do that only once because it is now an option for future projects saved in Revit's Imperial Library. To calculate the total amount of recycled content of the floors, you will need to add a calculated value.

Calculating the Volume Using the New Schedule

1. Click Calculated Value in the Schedule Properties window.

2. Enter **Total Recycled Volume** for the name.

3. Select the Formula setting.

4. Leave the default setting for Discipline.

5. In the Type drop-down list, select Volume.

6. Under Formula, enter **Volume*Recycled Content** (Figure 6.9).

7. Before exiting the Schedule Properties window, click OK until you are out of the Element Properties window.

8. Click Formatting, and select the field Recycled Content.

9. Now click Field Format.

10. Uncheck the default settings, and set Units to Percentage.

11. Set Rounding to 0 Decimal Places, and set Unit Symbol to the % sign, as shown in Figure 6.10.

12. Finally, click OK until you are back at the floor schedule.

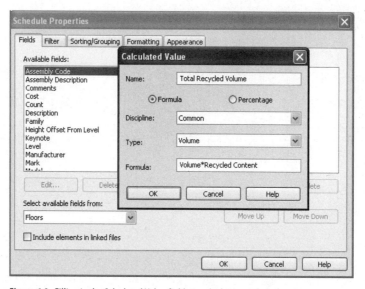

Figure 6.9 Filling in the Calculated Value fields to calculate recycled content

You should notice two new fields in your schedule: Recycled Content and Total Recycled Volume.

Click within the Recycled Content field, and enter **35** to indicate its recycled content is 35 percent (Figure 6.11). You now see that it changes the value to a percentage and in addition calculates your total recycled volume! This tool becomes invaluable, because as you move forward on the design of LEED projects, the model can change, and the schedule gets updated. The real advantage of your new field is that you can now see it under the available fields to use when you create a material schedule for doors, walls, ceilings, and so on. You can add rules and calculations and edit them to generate essentially any calculation rule necessary.

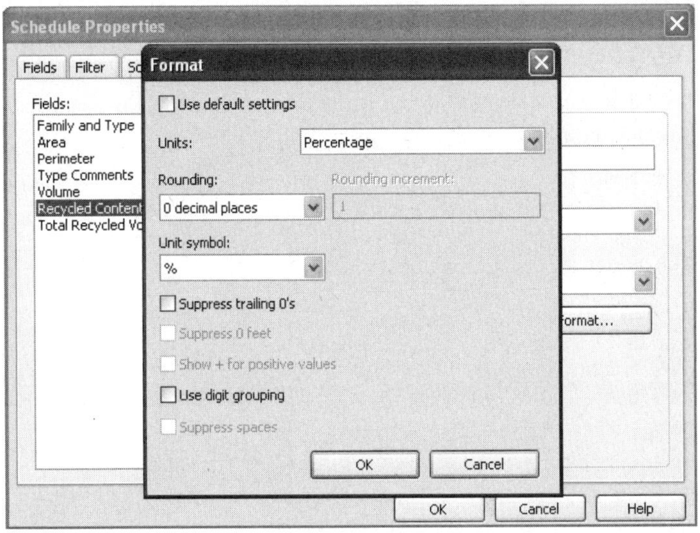

Figure 6.10 Sample image of completed calculated value

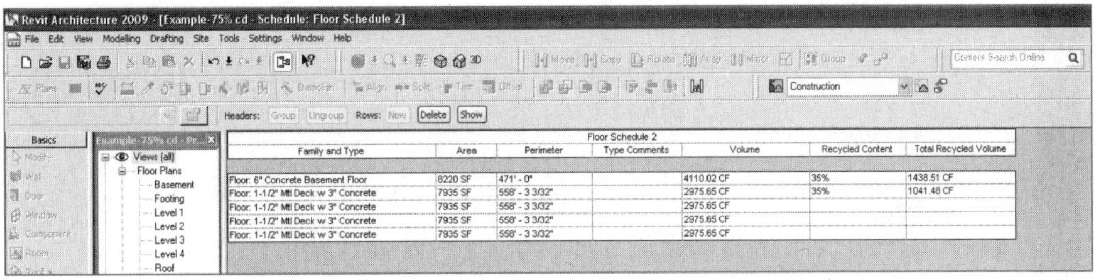

Figure 6.11 The completed schedule will automatically generate and update recycled material properties.

Material Radius

Material proximity is often a LEED requirement design teams strive for if they are trying to achieve a regionally sourced LEED credit. In LEED, these proximities are typically within a 500-mile radius of a project. The driving goals in this decision are to minimize shipping costs and pollution, promote locally manufactured goods and materials, and increase awareness about regional material resources. Holistically thinking about materials doesn't end if the product meets the recycled content criteria. What if that product comes from abroad for a project in the United States? Then what are the environmental impacts? For example, if an office building project in New York finds recycled content carpet that meets their criteria but the product is manufactured in China, a significant number of negative aspects typically more than offset the positive aspects of the materials. These include fuel consumption for the cargo ship carrying the carpet, airfreight or ground transportation pollution to the distributor, and sometimes even the packaging of multiple boxes. All of this amounts to a large amount of "embodied energy" into the product that would be much different if it were coming from New Jersey.

So, what tools are there to identify distances? Of course, some simple online resources allow you to type in going from point A to point B. But for a continuing resource, it would be nice to begin building a database of material suppliers' locations that allow you to check their radius from a project as project locations change. In the following tutorial, I'll show how to use the Google Earth tool to enter a specific location using the Placemark function to tell you how far from the project location a manufacturer is. In addition, this tool is often useful because new construction projects might not have an address yet or are in remote locations, and by using Google Earth to locate your building, you can tell whether a manufacturer is located within the required 500-mile radius. Google Earth also allows you to save the Placemark addresses of material suppliers that ease LEED reporting.

Material Radius Tutorial

To begin this tutorial, open Google Earth. For this example, you'll renovate the Space Needle in Seattle.

Creating Placemarks to Find Multiple Distances

1.   Under the Fly to Field, enter **Seattle Space Needle**, and click Begin Search (Figure 6.12).

2.   You now need to start creating a database of manufacturer locations, so right-click My Places, and select Add, and then Folder (Figure 6.13).

Figure 6.12 Entering a project location in Google Earth

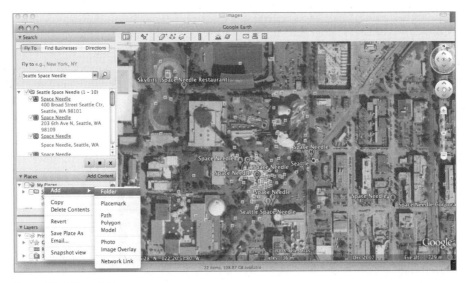

Figure 6.13 Adding a folder

3. Name the folder **Manufacturer Radius**.

4. Now right-click the new folder, and select the Add Placemark option.

5. Place the icon on the Space Needle, and label it **Jobsite** (Figure 6.14).

6. You will now enter two addresses of material suppliers. Enter the address **1303 SE 6th Ave Portland, Oregon** on the Fly To bar, and click Begin Search.

7. Add a Placemark here under the same Manufacturers folder, and name it **wood supplier** (Figure 6.15).

Figure 6.14 Naming the icon Jobsite

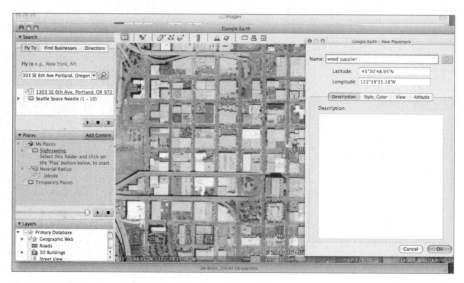

Figure 6.15 Adding a placemark to locate the wood supplier

8. Right-click the Places menu, where you created your new folder. Right-click the job site and select Directions from Here (Figure 6.16).

Figure 6.16 Finding the distance from the job site

9. Then, right-click the "wood supplier" placemark, and select Directions to Here. Google Earth automatically routes out the travel and summarizes the distance traveled.

In this case, the reclaimed wood supplier is approximately 176 miles from the job site, which is within the 500-mile radius and is acceptable (Figure 6.17).

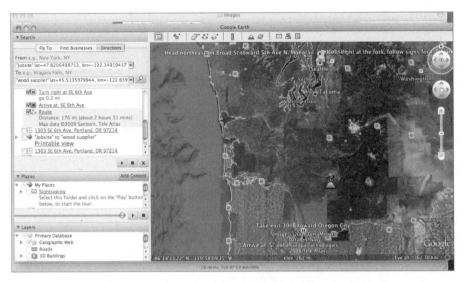

Figure 6.17 The Route tab identifies the total distance the supplier is located from the job site.

10. For our second example, enter the address **1482 Commerce Dr. Laramie, WY.**

11. Repeat steps 7–9, and name this Placemark **alternate wood supplier.**

What you find here is that the distance to the job site from this Wyoming mill is approximately 1,200 miles, which falls out of the 500-mile radius required to achieve a LEED credit (Figure 6.18).

Figure 6.18 The distance in this example exceeds the allowable 500-mile radius.

Essentially, by building a series of Placemarks for often-used or preferred material suppliers, a contractor can work with subcontractors to direct which suppliers are within a needed project radius. As the network of places increases, begin using the Description field for the placemarks to provide a description and contact information for the material provider. This approach limits the amount of rework you need to do for future calculations by not using a simple map tool and by being able to save the placemarks and project locations as necessary. You can also share the Google Earth placemark listings between tablet PCs and many handheld devices using the Add > Network Link function.

Lean Construction

Reducing material use has also become more popular as a means to cost effectively reduce the burden of landfills. The EPA estimates that 136 million tons of building-related construction and demolition (C&D) debris was generated in the United States in a single year ("Characterization of Construction and Demolition Debris in the United States, 1997 Update," www.epa.gov/epaoswer/non-hw/debris/about.htm). As shocking as this number is, we can lessen the amount of waste we generate from construction projects every year by more accurately ordering materials and decreasing unnecessary overage factors. Through teaming with architects and engineers, the ability to carry over "lean" manufacturing practices into the construction industry is a possibility.

The concept of *lean construction* is an end-to-end means of "a holistic pursuit of concurrent and continuous improvements in all dimensions of the built and natural environment: design, construction, activation, maintenance, salvaging, and recycling. This approach tries to manage and improve construction processes with minimum cost and maximum value by considering customer needs." According to the book *The Foundations of Lean Construction*, lean manufacturing was initially a concept developed by Toyota when the company noticed that the amount of material, time, and resources that it lost in the design to construction phases of their automobiles was almost 50 percent of the project cost. Toyota then developed a collaborative team approach starting at the initial design meeting. This involved *all* players, from the engineers who were designing the engine, wiring, and computer systems to the interior designers who were designing the buttons, knobs, dashes, and so on. And what Toyota saw was a huge increase in profitability (about 34 percent). I have heard that the typical project cost for a single car is somewhere in the neighborhood of $13 million, give or take. Needless to say, when profitability increased by about $4.5 million a car, Toyota integrated it as a company standard.

In addition to management savings, two more areas increased the bottom line at Toyota. The first was that Toyota began to notice as the teams worked together more that the amount of staff actually needed on each project was significantly less (almost half), because the teams started to refine the process of knowing what questions to ask their teammates. This freed up other personnel to assist in other efforts or projects and again drove up the profitability of the project because overhead was limited. The second area of savings was the amount of material. Toyota's own internal green effort was established to try to eliminate product waste. So, although the teams were designing, the manufacturers (in the meetings) of the materials were able to produce their product based on the design to save waste. The example I heard in a conference was that instead of using a 4×6 standard piece of vinyl door covering, they switched to a 3×4 piece, which worked just as well, covered the door panel, reduced cost, and almost eliminated waste. This type of thinking is fundamentally different from the thinking of years past where the engineers and architects all functioned independently of another. The interconnectedness of the construction process makes it a necessity to work together, and the potential of limiting construction waste is one of the areas that is amplified through early communication and technology.

Learning from the Fabricator

During a design-build, contractor-led project I was working on, we all learned from the fabricator. We had been working with the engineers' and architects' BIM files to coordinate our clash detection. Yet, when we reached the stage of the project that required us to coordinate the mechanical engineers' BIM file with the fabricators' BIM, we noticed some significant differences.

The fabricator had altered the runs of some of the ductwork. By combining the models into the composite model in Navisworks, we were able to tell where the fabricators' design altered from the engineers' design.

The unique result was that the fabricator had looked at the initial design provided by the engineer and created more efficient runs that shortened the overall length of ductwork on the project by almost 22 percent! We ran the clash detection on the fabricator's model and found that there were only a couple of minimal fixes that needed to be introduced into the design. The new composite BIM was then sent to the engineer to review what had been done. The engineer responded that everything seemed to work out fine with the exception of only a couple of necessary lengths that the engineer needed to maintain the recommended friction and flow rates.

This amount of collaboration lessened the amount of duct materials and saved the project money. By introducing the subcontractor and fabricator in the process, we were able to realize material cost savings and reduce the necessary field labor to install.

In essence, with everyone being at the table with the right tools, you can create less waste, better utilize materials, and create more sustainable projects.

Sustainable Construction

Green construction is a relatively new way of thinking about building. Although green construction methodologies continue to advance, we are seeing an overall rise in the way contractors are thinking about the energy, water, and resources they use to build. Many times the amount of energy used on a job site is a large portion of monthly overhead costs. Even simple solutions such as insulating job trailers to reduce heating and cooling loads, unplugging devices when not in use, and using energy efficient lighting on-site can add up to both a serious energy and cost savings. As the field of green construction grows, the area of green construction and BIM is almost a blank slate. So, how can BIM help during the construction process, and what technologies exist to better increase eco-friendly building practices? Within BIM is the ability to determine early on the strategies to increase the safety of people and eco-systems, to safely store green building materials, to use green building practices, and to increase efficiencies through technology.

Sustainable Construction Goal Setting

The ability to set sustainability goals early in a project, set recycling percentages, and limit delivery ranges to decrease fuel loads all are a change in process to follow a more integrated approach to design and construction. In a truly sustainable project, it is good to develop a sustainability plan early on in (LEED or no). This plan will delineate what unique work the contractor is responsible for and what duties the architect needs to fulfill. This plan should also outline critical deadlines for the project to begin any submittals and be very clear in identifying who needs to have a task done by a certain date.

A sustainability plan is typically a document that identifies the sustainability goals for a project and can consist of a number of items. Many times the sustainability goal is a version of the LEED checklist in which the contractor and design team agree on achievable project goals for that particular project type and then assign responsibilities. Other times the sustainability goals might be an overall document that contains local ordinances and initiatives, green building certification goals, and individual company goals that have been established. The sustainability plan should consist of the following:

- Project summary
- Accreditation goal summary
- All calculated goals (water, resource, and energy)
- Local C&D recycling resources
- Local municipal sustainability initiatives (for example, `www.seattle.gov/environment/building.htm`)

- Project limits (VOCs, construction waste, dust and debris management, and so on)
- Project initiatives (green energy credits, on-site energy management, and so on)
- Evaluation

Sustainable Construction and BIM: Masdar Headquarters Case Study

By Robert Forest, AIA, OAA, RIBA, LEED; Partner,
Adrian Smith + Gordon Gill Architecture (AS+GG)

Adrian Smith + Gordon Gill Architecture is still early in the design process with the Masdar Headquarters located in Abu Dhabi, United Arab Emirates. Yet we are already seeing the benefits of using BIM.

Specifically, we are using Autodesk Revit on this project. We have just begun the design development phase of work, and the BIM file has already allowed us to advance various sustainable aspects of the project beyond what we would have been able to do with a non-BIM process. BIM works hand in hand with our sustainable design philosophy, enhancing our ability to deliver sustainable projects. In turn, AS+GG is exploring BIM's potential in many different aspects/dimensions on the project.

The Masdar Headquarters project has forced us to question standard practice and look at things holistically and in general differently, with BIM as a major tool assisting us. For the Masdar Headquarters, the project involved there was an initial international design competition, which Adrian Smith + Gordon Gill Architecture won. Then, the fast-track design and construction process began. BIM was used initially in the competition and then implemented fully when concept design began. All documents are now being produced from the BIM file.

IMAGE COURTESY OF ADRIAN SMITH + GORDON GILL ARCHITECTURE

continues

Sustainable Construction and BIM: Masdar Headquarters Case Study
(continued)

Masdar Background

Masdar is an Arabic word for the *source*. Masdar City was conceived and designed to position Abu Dhabi as a global leader in the research and development of renewable energy technology. It will champion the adoption of advanced energy technologies and contribute to the diversification and sustainable growth of the Abu Dhabi economy.

Masdar City is six-square kilometers in area and can accommodate 50,000 people. It is intended to become an international benchmark of sustainable urban development showcasing state-of-the-art energy and waste efficiency, sustainable working environments, leading-edge research, and commercial development in renewable energy technologies, as well as a center for academic excellence. The Masdar Institute of Science and Technology is part of a joint venture with MIT.

Masdar City is intended to reduce Abu Dhabi's dependence on fossil fuels and reduce pollution, bringing Abu Dhabi and the UAE to the forefront of energy research and development, as well as branding Abu Dhabi as an environmental leader and model to the world.

Masdar is a city that is based on performance. Its goals are performance focused and quantifiable, albeit lofty, groundbreaking, and difficult to achieve.

Process

Initially when faced with the competition, we knew that an integrated design approach was imperative if we were to meet the Masdar City goals, which were the following:

- Zero carbon emissions
- Zero waste
- 100 percent power through renewable energy
- Zero accidents (through efficient health and safety protocols)
- Zero claims (effective PM systems and communication)
- Paperless document management system
- Energy savings and efficiency

We embraced these goals in all facets of our project execution and delivery and provided integrated services and methods that challenged conventional project solutions. We created an integrated consultant team with the major consultants, Structural (Thornton Tomasetti) and Environmental Systems Design (MEP), as well as our subconsultants working alongside Adrian Smith + Gordon Gill Architecture from the first conceptual discussions.

Sustainable Construction and BIM: Masdar Headquarters Case Study
(continued)

This integrated approach is allowing us to explore systems integration, be it structural, MEP, or others, with the architecture so as to increase the efficiency of all systems.

IMAGE COURTESY OF ADRIAN SMITH + GORDON GILL ARCHITECTURE

Using BIM is assisting in integrating the team and the process. Because each consultant has ownership of elements in the BIM file, the teams are working closer together on the model and document development. One team member's action directly relates to the others in real time. Coordination sessions are hourly, not weekly. The traditional thinking of "I will draw it the way it should be and then inform the other consultants to revise their drawings accordingly" is reversed, and as a result, efficiency therefore sustainability are enhanced.

The integrated design process allowed us to capitalize on the true value of sustainable opportunities at the beginning of the process. As the project proceeds through the design phases, where systems and design decisions are locked in, the sustainable opportunities and opportunity implementation costs increase exponentially. And, an integrated approach ensures the team is thinking about integrating the same concepts from the beginning and not waiting until it is traditionally "appropriate" to review and confirm the systems' performance.

BIM is allowing us to validate design assumption with certainty. The BIM is integrating with the other analysis programs that we need to validate our design assumptions. We have used the CFD software FloVENT from Mentor Graphics to study wind cone flow and additional CFD software to study the radiant temperature at the roof level under the PV trellis. The BIM file was exported to these programs, and its accuracy assisted our ability to predict conditions and revise the solutions if necessary. For example, when studying the wind cones and openings at the base, we found that creating openings that are diagonal to each other was more beneficial to the project overall because it created the desired wind flow throughout the cones. Openings that are orthogonal to each other created "flow through" with the wind passing through the cones and not creating as much desired wind flow as the diagonal openings.

continues

Sustainable Construction and BIM: Masdar Headquarters Case Study
(continued)

Sustainable Construction

AS+GG is dedicated to sustainable design and believes that the concept of that sustainability should not begin once the Masdar Headquarters' doors open. We questioned ourselves: "What if the design and construction process were so sustainable that a building would be 'carbon neutral' from its first day of construction?"

We looked at the construction process and reviewed ways to reduce the building's carbon footprint during construction and reviewed strategies to enhance construction efficiencies, thereby increasing the building's sustainability.

We rethought the construction process and proposed that the building's roof/trellis with PV panels and the supporting cones be built first to produce "clean" electricity during construction. The energy that cannot be used directly on-site is sent back to the grid, therefore creating a net-zero energy construction site and creating carbon credits to offset the construction activity's carbon emissions. We prepared the design so that this process could be implemented—we designed the cones, roof/trellis, and PVs to be independent of the concrete floor slabs. The BIM file assisted with the review of the complex geometry interfaces, which allowed us to ensure the construction strategy was possible.

IMAGE COURTESY OF ADRIAN SMITH + GORDON GILL ARCHITECTURE

Sustainable Construction and BIM: Masdar Headquarters Case Study

(continued)

In Abu Dhabi, construction stops during the afternoon in the summer months because of the extreme temperatures and risks to the construction workers. Another aspect of sustainable construction for the Masdar Headquarters is that our concept of providing the roof/trellis and PVs was also to provide shade and create a microclimate for the workers. The design of the trellis includes a large overhang, taking into account the sun position and ensuring the interior of the building is in shade during construction. The potential reduction in temperature locally will improve worker performance, thus enhancing the construction efficiency and therefore the overall sustainability of the project.

The BIM file was linked to the construction schedule to allow a 4D view of the project and an evaluation of the viability of two construction methods. One method was to construct the roof and PVs first as a more sustainable process, and the second was a more traditional construction sequence. Once analyzed, the traditional method was found to be quicker. Currently, the owner is evaluating both methods to determine the ultimate cost effectiveness, sustainable implications, and schedule impact. AS+GG is evaluating the traditional construction method with revised sustainable concepts to help the workers be more efficient (shade) and to make the project carbon neutral from other sources (other renewable energy solutions are a solar collector farm or temporary PV farm).

BIM has allowed us to take the architectural documents to another dimension. We were at 2D, and 3D was essential to illustrate the architectural space to owners. Now we are pushing the BIM to be 4D to include construction schedule information. Tying the schedule to the model and process is key in true sustainability. We are also looking beyond 4D to do the following:

- 5D: Adding cost/budget components.
- 6D: Adding facilities management components.
- 7D: Sustainability components (manufacturers, recycled content, embodied carbon, and so on).
- 8D: We are exploring the next suitable aspect.

Phases/Schedule

We used BIM very early in the process, implementing it fully in the beginning of concept design. This is difficult because basic decisions have not been made and the model information is still in flux, but we were able to successfully implement BIM early because the basic design had been accepted by the owner in the competition phase. This allowed us to begin to model from concept design, but it is not typical.

continues

Sustainable Construction and BIM: Masdar Headquarters Case Study
(continued)

BIM blurs the lines of the design phases as the team begins to work out basic details at the beginning of the process where you would have waited in the traditional method. Design development items get done in schematic design, and some schematic design items wait longer to be implemented. BIM has allowed us to issue for tender piling and steel packages tremendously early in the process, in the schematic design phase, thereby allowing us to carry out a super fast-track or "hypertrack" design process, as we call it. As soon as we draw an element such as a piece of the steelwork for the Masdar HQ cones, we can quantify it and prepare initial quantities.

Sustainable Content (7D information)

It is imperative to meet Masdar City's goals that recycled content and Embodied CO_2 are tracked with the BIM, thereby fully detailing the total building content. BIM has allowed us to set carbon targets for specific elements of the project and validate the design decisions accordingly or test different options and compare.

Once we have the data, we will also look at the total content and review the supply chain of products to enhance the recycled content. BIM allows us to have a real-time update on the recycled content status.

Economics

Our goal is to use the BIM file to assist with the cost of the project to give us a real-time view of costs and to assist in preparing accurate cost/budget models.

In addition, we rethought the concept of reforming the economic model for building and construction by introducing the sustainability aspect into the project. This added component brings value to the efficiency of the process as well as the project's life cycle.

IMAGE COURTESY OF ADRIAN SMITH + GORDON GILL ARCHITECTURE

Sustainable Construction and BIM: Masdar Headquarters Case Study
(continued)

Paperless

We have a paperless process with the BIM file at the center of the information with different parties using only computers and different aspects of the model as required.

City BIM Goal

One of the ultimate goals of Masdar City is to use the BIM file for the Masdar Headquarters project and integrate the future building projects of Masdar City to create a truly intelligent database on a city scale. This new tool will be a database that will track the carbon footprint and energy use for the entire city and will allow a "'real-time view" on the city's energy performance. This real-time view could be accessed by any resident to view their respective performance in relation to the city.

Future

We are starting to realize the potential of BIM with sustainable design/construction, but there is more to be accomplished. Innovative architects, designers, consultants, constructors, owners, and tenants will demand more, and as more potential is harnessed, there are more opportunities.

Overall, sustainability planning is critical to the success of a project, because during that time you see the project take shape and verify that goals are reached. A sustainability plan is an excellent resource for those who are newcomers to the team to quickly become acclimated to the project's initiatives and goals. In turn, this helps all parties become educated because all participants have a reference as they complete each scope of their work to relevant goals within an area. In addition, a sustainability plan is a great tool for an owner to see how the project is going to achieve its sustainability goals. Just as you use construction documentation and schedules to show information and move toward project completion, the sustainability plan is a similar resource. The sustainability plan is most effectively completed at the onset of a project. In fact, the project-planning period that includes developing a model coordination and information exchange plan will streamline tasks.

Job Site Surveillance

Job site surveillance, security, and safety have become more important than ever in recent years. And in an effort to further the tools available to construction managers, many companies now offer solar-powered job lighting teamed with security cameras. These cameras can be controlled off-site through a laptop or mobile device to scroll, pan, and zoom into the relevant area. On a job that involves a large number of remote managers, this technology could potentially limit the number of trips and fuel costs that a project manager would have to make. In addition, the use of handheld tablets teamed with a BIM file can serve as a means for a construction manager to be everywhere at once. This capability can also increase the superintendents' efficiencies because they are able to view in real time the issues on a project from anywhere—on the job site or on a different job. In addition, by using the BIM file, the superintendent can create a list of potential issues using Adobe Acrobat, Navisworks, or similar RFI software. Especially on larger projects such as stadiums, arenas, conference centers, casinos, hotels, and condominium projects that might involve multiple floors, a live camera feed that can be accessed through the Web is an efficient means of limiting travel time as well as cutting down on unnecessary travel by other project team members.

The Green Trailer

Never underestimate the ability to innovate. During my time at McCownGordon, we asked how we as a construction company could be more sustainable. We started to investigate the problem by asking these questions: How are we using energy? Where are we most inefficient? How can we make people rethink construction? What about our job trailers?

That's right—the job trailer. The answer surprised even us! Yet as we investigated this, we began to see the potential for a solution. The average construction job trailer runs 24/7 often with lights on and with little or no regard for energy efficiency. In fact, the average job trailer consumes enough electricity to power eight homes every month!

That decided, we set forward on innovating. The answer came in the form of the greenest trailer we'd ever seen:

- Using a hybrid solar- and wind-powered job trailer, the trailer was completely energy independent.

- By reusing an existing trailer frame, we were diverting material from the landfill.

- Using reclaimed lumber and clad with reclaimed wood inside, we all but erased our materials' embodied energy.

- The composting toilet used powerful bacteria and kept the trailer from using water or harmful chemicals to treat waste.

- The auxiliary fan system used natural ventilation to move air through the space.

The first time the trailer rolled onto a powerless job site and was up and running in a completely self-sustaining manner meant we were leading by example. In fact, we were energy positive as the workers plugged their power tools into the trailer until the temporary power was installed.

The Green Trailer *(continued)*

Using BIM, the trailer was designed for correct solar study and orientation, for reclaimed and recycled content percentages, and for construction documents. Later, the BIM file was used for graphics, and by having the BIM file already built, we could place it on a job site and orient it correctly as well. The trailer currently applies for a LEED innovation credit and are converting other job trailers into green trailers as well.

A new way of thinking about job site surveillance is from an environmental perspective. Green projects are often not constructed according to green standards and principles. Although part of this is because of a lack of understanding, some of this is attributed to improper site management. Today's technology provides construction managers with tools to inform themselves and direct their staffs to make sure a project is being built sustainably and correctly. It is believed that, much like Jeremy Bentham's idea of the panopticon, the concept of being watched increases your attention to details and decreases inappropriate behavior such as illegal dumping, wetland disturbance, and other logistical issues that might arise on a site. However, you also need to understand site weather conditions and potentially hazardous areas. In the future, the ability for a project manager to communicate and locate personnel through GPS on the job site or by sending live web links of video feeds to superintendents and foremen are all very much possibilities.

Material Management

Material waste as a result of having to trash damaged materials is not only costly but also a very wasteful way of managing material resources. This sort of waste is not acceptable when you look at the technology available to prevent site waste and to manage on-site deliveries. Effectively managing materials is also critical when working on a green project because waste factors generated on-site can actually reduce the amount of LEED credits a project is capable of earning at its completion. So, how can you utilize technology to better manage site material storage and condition?

The first answer is through effective means of scheduling. Earlier you saw an example of using the BIM file to show when a portion of a project is to begin and then to be completed using Navisworks. This same type of methodology can directly carry over into the prompt and timely delivery of construction materials. The fewer materials that are on-site, the less chance those materials have of becoming damaged before their use. However, this also requires a balance of having the right materials on-site to make sure that progress isn't being halted because of a lack of materials. To manage this process, construction managers can use a phasing plan in the form of material orders that are purposed for each portion of the project.

Project phasing models can be built to show the anticipated delivery date of materials on a job site either through animations or through a series of stills. Using Revit or other modeling software and Navisworks, you can create simple masses that represent piles of CMU blocks or 2×4s to staging areas for the structural steel and precast components on a project. Although this may seem somewhat redundant to the logistics plan, you can utilize the existing phasing animation and verify that large lengths of steel will fit in their specified staging areas or that multiple precast components may be delivered at the same time without known site conflicts that could have been avoided, just as you created the phasing animation.

Another BIM tool for sustainable material management is accurate material quantities. For example, if a hotel project is dried in and the construction is finishing in a top-down manner, then the construction manager can utilize the BIM file to quickly estimate what quantities of materials will be needed for each floor. As the project progresses, the construction manager can quickly gain an accurate understanding of the required materials for each floor, and by the time the finish out reaches the ground, the construction manager has streamlined the material ordering and coordination enough to know what will be needed on each floor. BIM can also be used to quantify one-time material needs for nonrepetitive work such as a custom loft or museum. Using the material schedule in Revit or the quantity manager in Innovaya, a construction manager can get an accurate representation of the amount of material needed for a particular area of work provided the model is accurate.

To successfully use the BIM file for in-field material ordering, the BIM file needs to be to a relatively high level of model completion (not detail or drafting completion). Although this remains true for all the tools mentioned in this book, it is particularly true as the profession moves forward and as BIM users become more efficient and accurate in the amount of information they are inputting into the models.

Salvaging and Recycling

Green construction involves the end of the building's life cycle. The term *life cycle* refers to the concept that a building has a limited life span. Generally speaking, buildings aren't meant to last forever. Unfortunately, buildings often outlive their technology to the point where the systems that exist in a structure are outdated, material research is completed (asbestos, for example), and the environment around a structure changes in density, use, or purpose. Although many see these structures as landfill material, often that is not the case. In fact, many of the materials in a structure can be repurposed with a minimal investment of time and natural resources.

Salvaging

Salvage is, of course, the greenest way to reintroduce materials into the building product stream (Figure 6.19 and Figure 6.20). This is because of the embodied energy already contained within a building material. For example, a brick that is reclaimed from a job site, cleaned, stacked, and delivered to a new project has very little embodied energy when compared to a new brick. The reclaimed brick does not have the energy of mining the clay, forming, kiln baking, and curing that a new brick has. The future of green building reclamation has exploded in recent years and will probably continue at a staggering pace. Much of this growth can be attributed to rising material costs, high fuel prices, and an overall lack of material. Constantly increasing steel prices is a good example. Organizations such as the Building Materials Reuse Association,

PlanetReuse, and other material reclamation companies facilitate the sustainable reuse of building materials. Although this market is driven by economics, the fact remains that economics will continue to drive the market to reuse because demolition contractors are now seeing a pile of potential profit instead of a pile of debris. So, how can we use BIM to help us know what is salvageable?

Figure 6.19 Deconstructed salvaged brick from a job site

Figure 6.20 The same salvaged brick from a job site ready for reuse

To start, we know that there is a huge value in inserting information about material into the model. Yet the manufacturing industry as a whole has been slow to fully embrace BIM technology. Currently, this is because of a lack of understanding and education as to what BIM is. Some believe that a BIM file is anything in 3D sans intelligence such as a Sketchup or 3D Studio Max model, and, yes, through a series of importing and adding information, you can add parameters to a component to make this model work. However, most architects and engineers don't have the time to spend customizing every family and assembly because the products they are using vary just as the projects they are designing vary. Currently, manufacturers are missing an enormous opportunity within the AEC community to build or coordinate online libraries of BIM family content. In doing this, manufacturers would be recognized as a resource and have their product used as the basis for design. Ideally, this would mean that for a facility nearing the end of its life cycle, the facility manager could extract from the BIM all of the building's components, manufacturers, quantities, and recipes of assembly and deliver these to the demolition or deconstruction contractor. The deconstruction contractor would then have a tool to quantify the materials in a structure that are to be decommissioned and could begin finding resources for these materials prior to the structure being dismantled. The future of deconstruction lies in being able to access the information in a facility to limit expense, find sources for materials, and make better informed decisions when it comes to estimating the cost to deconstruct facilities. In fact, according to the Green Building Alliance, the green building product market is projected to be worth $30 to $40 billion annually by 2010. The industry will continue to find ways to "close the loop," and because they will do this, we can plan for it now and encourage manufacturers to provide additional data and tools.

So, how do we make it work now? Well, embedding URL links into a Revit file is possible. Yet these run the potential risk of a website being changed or deleted and therefore any information linking to that site then unreferenced. You can input the information yourself, yet the business case for an undertaking such as this would have to be developed project by project with very little ROI from the project team, though it might make sense for the facility manager. That said, there is currently a gap in the industry in regard to the availability of model components that contain even basic information about their building products. Ideally, it would be good to know who manufactured the material, what those products are made of (latex, synthetics, wood, fiber, plastics), how environmentally sound those material are (or a Pharos rating, www.pharos .net), recycled content, material location, and lastly how much of it there is. In essence, a nutritional label of sorts that would add a much needed layer of information about a project would be a terrific resource for the AECO industry. Decisions could be made based on the degree of sustainability in a product's model component. Moreover, the idea is for users to have the opportunity to insert this information and need to work with partners to encourage them to move BIM technology forward.

How can you further promote material BIM libraries? The simple fact that architecture and construction firms find it difficult to invest in creating and updating a library that might contain a large amount of the information about a building material within the model makes for a difficult solution. Yet, as the industry continues to move forward, I believe we will see more and more manufacturers not only get on board with BIM but also champion sustainability efforts in the form of online BIM libraries that contain accurate and comprehensive amounts of data about their products. Essentially, this is an area that can vastly be improved, and with the ability for software such as e-SPECS and other BIM specification–generating software, it is easy to assume that there is value to manufacturers, both from the standpoint of marketing exposure and from a green standpoint by keeping their material libraries online as opposed to print material. In the future, the construction industry will begin to understand all that BIM has to offer and start providing valuable feedback to manufacturers about the tools they need to provide. Just as CAD resources are available online now from a number of manufacturers, you should expect to see the same tools made available for BIM in the near future.

Recycling

Where material reclamation isn't a possibility, material recycling is becoming an industry standard, not only as a standard for buildings coming down but also for building material excess during construction. There are many projects with initiatives to be zero waste, which of course is the most sustainable option. Yet, when this might not be possible (for a variety of reasons), on-site recycling is a necessity. Using BIM, you can provide a number of tools in regard to job site recycling. First, using the site logistical plan, you can identify where bins and rolloffs are located on the site. Using this in initial project construction meetings, introduce the team to these locations, and if there are project credit goals regarding these items, then it becomes very important that subcontractors know how to take care of scraps. The second way you can utilize BIM to improve recycling is to limit the amount of printed documentation on-site. Through the use of on-site tablet PCs, an outfitted job trailer, or digital document coordination, you are again reducing the job's footprint. Although paper might seem like a small portion of waste being generated from a project, keep in mind that construction documentation in the form of specifications can range anywhere from 100 to 10,000 pages. Multiply this by the number of team members and stakeholders, and you'll see that a significant amount of paper is changing hands. In addition, multiply the cost of required copies to be distributed among those who are bidding a project, and you run into serious expense as well. Although BIM is not perfect and may not completely eliminate all paper on a job site, it can definitely reduce the amount of printed documentation needed and start making users get into the habit of referring to digital tools as opposed to information printed on a natural resource.

Conclusion

In William McDonough's book *Cradle to Cradle*, he lays a strong framework for a philosophy of rethinking the way we make and construct things. And by *things*, I mean literally everything. In the way we choose materials, we need to think about how the material can be reused and what part (or all) of it can be recycled at the end of the product's life. In the way we use fuel, we need to think about how to consume limited resources and how to use materials that can be reused or recycled, with nothing going to the landfill. As more contractors use cutting-edge technology in BIM, we should be able to use the information to make more informed decisions and, in addition, steer the manufacturers of everything from the brick the masons lay to the concrete truck used. Sustainable construction, BIM, and efficient design all go hand in hand as we continue to find ways to equip future generations with the tools to achieve what we currently can't, and to build as sustainably as we can as we borrow the Earth from future generations.

BIM and Facility Management

In this chapter, I will cover how to use BIM during the facility management phase of a structure. Specifically, I'll cover the following topics:

Costs of Maintaining a Facility

A building's operating cost over its life cycle is on average approximately 80 percent of the total initial cost to build and far outweighs the initial 20 percent investment to construct it (Figure 7.1). Yet, a building's operating costs and management are regarded as necessary, not areas that could provide potential savings. Additionally, the high environmental costs associated with reducing pollution emissions and vacancy costs associated with poor overall occupant health are additional costs that face today's facility owner. In fact, more than 80 percent of facility owners in a recent study found that owners are budgeting for green initiatives and energy savings in their facilities (source: www.facilitiesnet.com/news/article.asp?id=10192) to reduce these additional factors. It's interesting in the current construction industry just how much of the focus is on the first cost of design and construction as opposed to the enduring cost of facility operations. The opportunities to design buildings to be more efficient, build them to healthier standards, and maintain them to the best operating standard is becoming more critical in today's economy. Especially if you consider the impact even a small annual savings would have when multiplied by the years the building is in use.

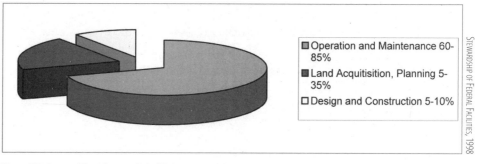

Operation and Maintenance 60-85%

Land Acquitisition, Planning 5-35%

Design and Construction 5-10%

STEWARDSHIP OF FEDERAL FACILITIES, 1998

Figure 7.1 Average life cycle cost of a building

Rising energy costs, material replacement, and tenant turnover are just some of the challenges a building owner faces. Facility managers and building owners alike are finding that, in order to be competitive with lease rates and capital allocation, they must have a highly efficient, healthy, and well maintained facility. A recent survey of facility managers and property owners around the world (completed by Johnson Controls) found that the anticipated costs of operating a facility have risen approximately 13 percent over the past two years and will continue to rise in the future (source: www.fmlink.com). Not only are energy costs rising, but there is an increased demand within the market for these healthier and more sustainable buildings from building tenants. Although this requirement doesn't reflect any direct cost, it does translate into marketability. New sustainable construction and retrofitting has increased drastically, and 88 percent of building owners have said that energy efficiency was a design priority in their construction and retrofit projects (source: www.fmlink.com).

In addition, building owners are challenged with the repairs and cleaning costs for an operation, which can add up to $2 to $5 per square foot for the life of the facility (source: www.fmlink.com). To put this into perspective, for a 40,000-square-foot building, the average cost would be approximately $200,000 a year. Multiplied by a 30-year building life, the total cost of maintenance is more than $6 million (not considering inflation). More important, these are overhead costs, with nothing going toward the cost of operating the building yet! For example sake, the same new 40,000-square-foot office building costs somewhere around $185 per square foot. This building would cost about $7.4 million to build (Figure 7.2). Contrasting the cost of construction with the cost of minimal cleaning and repairs, the result is almost equal.

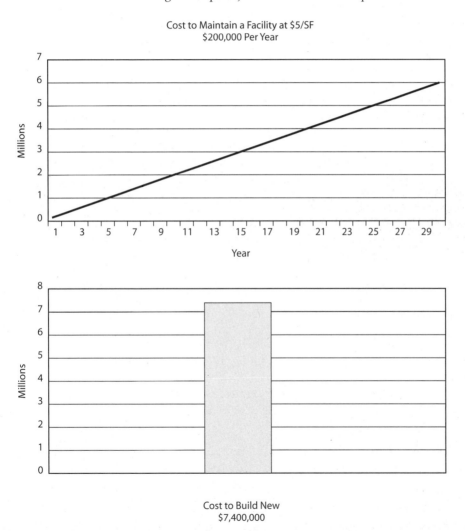

Figure 7.2 The costs of a building over its life cycle almost equate to its initial construction costs.

Additionally, there are other costs to tack on this, including inefficiencies in response time, asset loss, facility management staff, and the transfer of data between software systems. According to David Jordani, construction is a $3 trillion industry with 50 percent waste generated through its life cycle (source: www.specialtypub.com). This number equates to a serious loss of profit for building owners and, as a result, is reflected in the time and quality of a construction project as well as its operation over its life cycle.

In the past, CAD files were distributed to the facility manager upon project completion. Paid for by the owner, the files were included as part of the contract language from the architect and construction manager. These documents then became the sole means of reference for the operations of the facility. The cost to procure these documents varied but was generally a relatively small cost when compared to the other portions of the AEC team's fee. In addition, these CAD files weren't ready for the facility manager to use. To get the drawings to feed information to the majority of CAD software, you would need to have the rooms traced or polylined. (*Polylines* are closed loops composed of line and arc segments that are used to define a room and area so that information can be assigned to it.) CAD is not in 3D and as such doesn't define the rooms automatically by defining its three-dimensional boundaries like BIM does. This has resulted in an entire cottage polyline service industry that provides these services to facility managers for a cost. Software such as Autodesk FMDesktop offers some solutions to using BIM in space management tools that export the spatial data from the BIM database to a spreadsheet format and allow additional information to be linked to the database. Using these tools is great for room and spatial information as well as an example of using someone else's work in a BIM process to lessen the amount of work required.

The following is an example of information that can be included in a BIM file or linked to the BIM via a URL and/or an external or internal server:

- Architectural program
- Space functions
- Secure areas
- Area calculations
- Volume calculations
- Engineering performance criteria
- Specifications
- Contracts
- Survey information
- Change orders (potential and actual)
- Shop drawings and submittals
- Procurement files

- Progress photos
- Head wall photos
- Electronic diagrams
- Warranty information
- Invoices
- Purchase requests
- Work requests
- Estimates for work
- Occupant organization
- Seating arrangements
- Network diagrams
- Hazmat identification
- O and M manuals
- Inspection reports
- Commissioning reports
- Analysis reports and simulations
- Disaster recovery plans
- Asset management and tracking

Owners and BIM

Owners can benefit greatly through the use of an as-built model as a project deliverable. Currently, one of the largest problems in dealing with end-of-project information is the huge amount of documentation owners are left to deal with. For example, submittals provide very specific information about a product on everything from door hardware to HVAC systems and include extensive data about equipment such as the manufacturer, warranty, and color. The larger and more complex the project, the more submittals and thus the more information owners have to deal with. Specifications are another example of end-of-project data. Specifications call out the products used, the warranty requirements, and the installation instructions. This can mean huge amounts of unlinked information going through many hands, which can result in unorganized or, even worse, lost data. Some of the biggest costs associated with this system of data management is lost time in trying to locate manufacturer information for component repair or replacement. The time it takes facility managers to locate information on a particular part is proportional to their ability to meet the facility's other needs. Lost time searching for information sets the facility manager that much further behind in properly maintaining the facility. In some cases, the warranties are even voided on HVAC, roofing, and flooring systems because of improper maintenance. This can result in the complete replacement of systems and shutting down or closing a facility, which isn't a feasible solution in some structures such as certain hospitals and high-security government buildings. Owners stand to save considerable dollars in a BIM process that result in a record BIM deliverable for the following reasons:

- Increase in staff efficiency to get to information (time)
- Keeping equipment maintained to warranty standards
- Properly documenting commissioning issues (life safety, fire stopping, accessibility)
- Limited amount of wasteful printing (costs)
- Ability to back up critical digital and facility data that could be lost
- Warranties embedded in the BIM and accessible to avoid unnecessary costs
- Less chance for facility downtime as a result of improper maintenance
- More efficient repair response
- Improved client/occupant satisfaction

A record BIM file, similar to other facility investments such as solar panels, sun shades, and high efficiency HVAC units, is a cost that an owner will realize over time. In addition, BIM provides a fingerprint of the facility over its lifecycle. This means significant savings in information finding and analysis, as well as providing design teams with accurate as-built information for renovations or additions to the facility. So, the question now becomes, why wouldn't an owner want a BIM delivery at project close-out if it better equipped the operations team?

Some believe that the costs associated with developing a record BIM should be included as part of the design teams deliverable. Others believe that there should be an increased fee for delivery and coordination of the record BIM. This scenario often leads designers and owners to question whether there should be an increased fee for this new product. The word *product* is intentional here because the definition of product according to the Random House Dictionary coincides directly with the BIM both as a thing produced by labor or a thing produced by or resulting from a process. What's interesting is that comparing a BIM deliverable to the CAD deliverable speaks volumes about the process that is in place for both technologies and their respective end results. The quality of the product reflects the quality of the process. Simply put, the owner is now receiving a better product in the record BIM, and the additional time and expense required to compile it should be reimbursed accordingly. Just as a consumer would pay more for a luxury car with leather seats and all of the bells and whistles than for an economy compact vehicle, the same is true in regard to delivering the record BIM, with the exception that the luxury vehicle includes an owner's manual! However, this doesn't necessarily mean that this is an area where a design team can significantly recapture lost profits or that the owner is willing to pay any extra for it. Although the delivery of the record BIM and additional costs associated with it still are being debated; it should be talked about as a fair exchange between both parties as the owner is receiving a better deliverable and the modeling team is taking on more responsibilities.

The fact that the facility management team could be given a much more usable and comprehensive tool could mean direct savings as well (Figure 7.3). Of course, it is the decision of the owner to require a record BIM as well as specify what components are needed in the model. Unless part of a business development or marketing effort, creating a record BIM is not considered a standard deliverable in current processes. On the contrary, the process of creating a record BIM is a markedly larger undertaking than in a CAD delivery at project closeout. In addition, creating an as-built record BIM file requires the AEC team on a project to coordinate much more than before in preparing the specifications, models, and field information because they will be used beyond design and construction. (See the "If I Had Known That, I Would Have Written the Specs Better!" sidebar.)

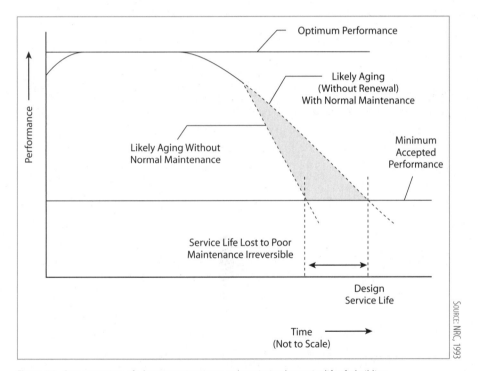

Figure 7.3 Cost comparison of adequate maintenance and repairs on the service life of a building

The delivery of a record BIM must be discussed during the preconstruction phase of the project. Some organizations and owners educated on BIM may require certain levels of detail within their models. An example of such a requirement is the Government Services Administration's "3D–4D Building Information Guide." According to the Government Services Administration (GSA), "The power of visualization, coordination, simulation, and optimization from 3D, 4D, and BIM computer technologies allow GSA to more effectively meet customer, design, construction, and program requirements. GSA is committed to a strategic and incremental adoption of 3D, 4D, and BIM technologies" (source: www.gsa.gov/bim). Other organizations outlining the use of BIM on their projects are the National Institutes of Health's move to Revit deliverables and the Veterans Administration's move to an IFC-based BIM deliverable (source: www.nibs.org).

Record BIMs

Through this book I have frequently made reference to the record BIM. Conceptually, a record BIM is the single composite reference model for a construction project that contains the information from the project embedded within the model and linked to all relevant files. Although this is possible using today's technology, it is hardly feasible using current processes. The process of completing a record BIM involves additional resources throughout construction. Resources such as subs and fabricators shouldn't hinder the development of the project but rather document the changes and actual construction of the project as agreed upon by the project team. Owners have begun to understand the value of a "complete" and accurate BIM as it relates to both construction and maintenance of their new or renovated facility, and many are willing to pay extra for this added value. In fact, many owners are contracting to have models built of their existing facilities without the involvement of any physical construction. A good example of this is the GSA's 3D–4D BIM solicitations. This is because BIM can provide a usable, linked source of documentation for a building throughout its lifecycle (source: www.gsa.gov/bim).

As I discussed in Chapter 2, the existing delivery methods include providing the facility manager with CAD documentation to use in managing the facility. Depending on building size, the team to manage the operations and maintenance of a facility will vary in numbers. What is important to note is the lack of flexibility within CAD information. Facility managers are often responsible for maintenance, event scheduling, setup and cleanup of the building, tenant management including move management, and data management for a building. What often happens is that facility managers, because of the physical demands of a building, rarely keep up with the digital documentation that is associated with it. This is because CAD requires multiple files to be managed to properly maintain the data associated with the building. As tenants move, repairs are made, facilities are upgraded, equipment is switched out, and facilities change use, the management of the data is the facility manager's alone. Typically, there is no involvement by an architect, engineer, or general contractor to document the facility changes that take place.

For a number of reasons, CAD is difficult to use for facility management. One reason is because of the quantity of files to manage. CAD files can easily run into the hundreds or even thousands depending on the project. CAD drawings are separate files that can be linked into a parent view for reference to manage the facility. Also, the facility manager will need CAD software and a level of ability and experience to operate the software. What works about documentation in CAD is the nature of the native format. For the most part, when the delivered product is CAD, all the files are in CAD and are able to be opened, reviewed, altered, and saved. However, the tasks of changing and updating are quite time-consuming. In BIM, a facility manager can update wall information, ceiling or lighting layouts, or floor finishes, and it equals a change throughout. In CAD, a change in one place does not equal a change in another; it equates to sorting through all the CAD drawings associated with the change and updating them individually. This is because of the parametric abilities within BIM software as opposed to the vector line–based drawings in CAD. The CAD files used by the facility manager often become only the plan representation of the facility, and the other drawings go by the wayside. Another downside to CAD is the lack of intelligence within the files. For example, in BIM, a wall will contain all the unique characteristics that define it—such as what it is made of, its height, its area, its volume, and so on—which can help a facility manager quickly drill down to the information they are seeking. By contrast, in CAD, the walls are represented by lines, and the amount of intelligence associated with them is information such as colors, layers, and plot weights. Thus, CAD must be used in tandem with specification information, operations and maintenance manuals, and the construction documents. Because this is not as efficient a use of the facility manager's time, traditional 2D CAD limits the ability of the facility manger to achieve other tasks when compared to BIM.

The last hurdle in CAD as it relates to facility operations is the accuracy of the CAD files. During a CAD process, field changes, minor alterations, and other changes are rarely documented in the CAD files. Often, creating true, as-built CAD data is not part of the architect's or engineer's scope of work, and as a result, the information that the facility manager gets from the beginning of the facility's life cycle is most likely dated and inaccurate.

So, how do you go about creating a record BIM? There are two ways of documenting a record BIM. In both, the terms and expected level of detail should be discussed during the contract negotiation phase to eliminate issues as the project nears completion. The first option and often the most feasible is to have the general contractor create the record BIM for the project. The general contractor has a significant advantage in managing and updating the model, because field personnel and support staff are dedicated full-time to constructing the project and as a result are constantly informed of the changes and construction issues throughout the course of construction on-site. If the general contractor creates the record BIM, the ownership of the model transfers to the contractor as the project is being built. Figure 7.4 shows an example record BIM. Measures can be put in place that allow the architect to review the model updates and approve them as needed to limit expense. This often creates a comfort level with creating a record BIM and limits the resources the architectural team needs to dedicate to the construction administration phase of a project.

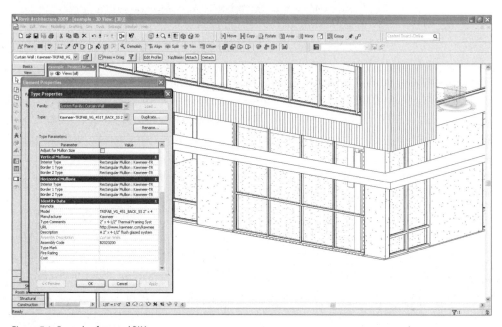

Figure 7.4 Example of a record BIM

The second way a BIM can be created is as part of a design or virtual construction professional's contract. This arrangement typically means that the architect must negotiate for more time and thus for more fee in the construction administration phase of a project. The responsibility to create a record BIM shifts as the project is being constructed, which means the model manager will need to have an intimate understanding of and involvement in the construction of the facility. The quality of what is being constructed must be reflected in the model to be effective. A fundamental shift in the way the industry is headed has begun with the change in thinking that "cheaper is better." This began with the rise of design-build and continues to build steam through other project delivery methods. The same trend has continued into the selection of the architecture and engineering teams; not only is cost considered now, but qualifications and value over the entire cost of the project are too, and this includes the ability to deliver a record BIM at the end of a project. By acquiring more of the responsibility in a project, the architect has the potential to capture more profit, while creating a usable tool that affects the way their building is operated.

The end result of a record BIM is to understand that virtual modeling and construction are not the old way of delivering "design intent" documents. Nor is it the contractor's responsibility to construct a design according to "means and methods." The reality is that BIM is becoming a fabrication model for contractors and an accurate as-built product. In his book update *Virtual Design and Construction: New Opportunities for Leadership* (*The Architect's Handbook of Professional Practice*, 2006), James R. Bedrick identifies the tremendous opportunity of the entire team to better collaborate and leverage BIM as a source of additional scope, responsibility, and potentially money.

Not only is BIM a faster way of documenting a project, but when fabricating certain components, the facility is ultimately constructed as modeled. The result of collaboration and accurate and information-rich modeling is a usable, virtual construction of a facility that enables the facility management team to better understand and maintain their structure.

One BIM = One Source of Information

Information in BIM is both visual and database driven. The concept of linking the visual representation with the spreadsheet, quantity, or other data source it's associated with is being pushed further into the realm of open-ended systems. The concept of open source programming or interoperability is part of the driving force behind the National Institute of Building Sciences (NIBS) and other organizations' efforts to allow software to talk with each other and for software companies to allow for customization and application development. This issue relates to BIM during facility management because the ability to manage and create logs, schedules, and other important resources means the facility manager can tie into the model to add layers of information on top of the model components. Although many programs and software companies claim they are driving toward an open source concept of information sharing in the form of open programming languages such as XML, GBXML and HTML, there is still a significant amount of improvement left in the field of BIM and facility management. Programs—such as Onuma Planning System (OPS), Archibus, and FMDesktop—have seen the opportunity in BIM and allow for the extraction of the room or spatial information by exporting the schedules. Although spatial validation/management is important, we have barely scratched the surface in regard to creating an information-rich source of data that can be freely linked to the model for use throughout the life of the building (Figure 7.5).

The question then becomes, how do we gather and effectively represent data and allow the facility manager to build upon this data for the life cycle of the building? The first solution is by creating a solid and accurate three-dimensional representation to begin with. Whether created by the contractor, architect, or virtual construction manager, the ability to utilize the model stems from the accuracy of its as-built representation in 3D. Having the correct information to start makes the facility manager's job much easier as adding information to correct model components is easier than editing model components *and* adding information. The approach shown in this book is one that uses the team member's model and then advances that file by using the model as reference to develop their information upon the other model. Ideally, models would be developed in a process where the model is created and work by other team members is directly input into the same file. This could really introduce a number of efficiencies and significant savings; however current limitations between software and model transferring do not allow this to happen.

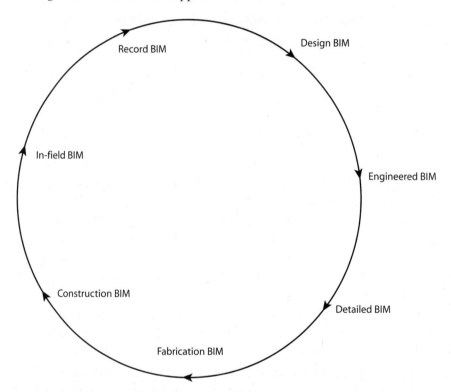

Figure 7.5 Record BIM deliverable process

Just as BIM is a process of defining goals and end results, so is the record BIM creation at preoperations. In reality, the record BIM will become a more important tool during operations because it will determine the efficiencies of the facility manager's team, the energy use of the building, and potentially even the health of its occupants over its life. As I discussed earlier, these costs far outweigh those initial upfront costs during construction and may equal significant savings to the owner. The last solution is to design, build, and document the facilities as if you were the operations manager.

The functions that a facility manager provides in general are managing the physical tasks associated with the proper operation of a building. This includes making sure all the equipment is being maintained, warranties are being documented, filters are being changed, and moves are being managed. Overall, a number of other tasks fall on the facility manager's shoulders such as properly documenting and resolving work requests, commissioning reports, and supplying architects and contractors with existing facility documentation for renovations, additions, or sizable repairs. This last responsibility is a difficult one to manage and often results in incorrect documentation across the board in a CAD process.

"If I Had Known That, I Would Have Written the Specs Better!"

Working on BIM projects often means that you learn about other team members just as much as you learn about your own profession. For example, I was working on a BIM project that was for a government owner group, and the contract specified that all design, construction, and project closeout deliverables were to be in BIM. The project went smoothly, and both the design and engineering teams were very accommodating in model sharing. Overall, the project went very smoothly...until project closeout.

A month before project closeout, we realized for the first time that we were to deliver a record BIM of the project. Aside from the formal construction documentation (RFI, ASD, ASI, and so on), we had done little to link and embed information within the model to meet the progressive BIM standard set forth by the owner, which we helped develop! So, we set to work and began linking the specifications, PDFs, and other information into the BIM file. While working on producing the final product, we all came to realize just how poorly we had created the documentation that the operations manager was to receive. Although this was no different from the standard level of information that other owners had come to accept, we found what we were delivering needed bridges over serious information gaps—so much so that one of the architects responded to warranty information on an operable partition with this: "If I had known we were supposed to produce this at the end of the project, I would have written the specs better!"

We learned from that experience that lots of information doesn't necessarily mean good information. It takes time to build the tool correctly and enable other downstream users to succeed.

A result of incomplete or inaccurate as-built data is the incurred cost to a building owner to pay for as-built research and documentation costs. This scope of this documentation phase can vary depending on the amount of time it takes to model/draft, field verify, or convert existing 2D documentation into usable information for a new project. In some cases, this may even mean closing part or all of a facility so that all of the information including the mechanical, electrical, and plumbing connections can be documented.

In a BIM process, the facility manager has the potential to maximize his efforts using the software. Although this may require training and the involvement of consultants at the front end of the investment, it could very likely pay for itself as the first renovation is scheduled and the as-built costs are diverted, if the facility manager has kept the information up-to-date. In addition, this will mean that the facility manager is building upon a tool that is making his day-to-day tasks easier. For example, after the third replacement of a VAV fan box from a different manufacturer with a different model number, the manager can quickly access the latest information about that part, saving time, preventing inaccurate replacement parts, and locate the associated warranty information that might mean no replacement costs at all.

The value of a single source of information is particularly relevant in life safety issues. The fire safety and emergency plans should reflect accurate floor plan layouts. In addition, the model should show where life safety devices such as water, gas, and fire shutoffs are located so that first responders can be most efficient. In fact, there are some cities that are investigating the use of BIM to make their ability to access building emergency equipment faster. The possibility of locating shutoffs, fire hose cabinets, extinguishers, and egress information while an emergency team is en route is an exciting proposition and one that could equate to saving a life. The scary reality is that facilities may have altered or outdated information that is incorrectly reflected in the life safety plans. Although this may merely complicate an emergency situation, it could also result in loss of life if outdated escape information is displayed in the facility.

Putting BIM as a resource in place for a facility manager decreases the amount of time it takes to get to and add information. BIM is still a tool that is only as valuable as the accuracy of the data input and the sophistication of the person using it. For this purpose, the process of managing facilities and what is to be expected from the construction and design teams at project closeout will continue to change. Educated owners will continue to demand more resources to equip their staff with the best tools to handle the job. These same owners have begun to hire from within the AEC community to better manage the information associated with their facilities. Conversely, more contractors who aren't capable of producing this delivery at the close of a project will be less competitive. BIM is effective as a single source of information through 3D representation and will continue to be further refined and developed by what and how operations personnel use it throughout its evolution.

Information in a Record BIM

How much information is too much information in a BIM file? This is often the question of not only the contractor but also of the architect, engineers, owner, and consultants as well. If you're enabling BIM through an integrated process, you have the ability to get the stakeholders to answer this question. The owner will let you know what is expected, and the facility management team will be able to voice their needs and tell you what they will be using the information for. The architects can communicate to the team and provide the level of information that they envision providing. All the while, the collaboration and data standards are being created, and the rest of the team can discuss what additional information and data sharing needs there might be. The beauty of this process is that the team has its own answers. Although this may seem like a bold statement, it's the truth. Projects may vary in expectation as well as size and complexity and even different team members, but the ability to have questions answered within the group is one of the main benefits to integrating teams between multiple stakeholders (Figure 7.6).

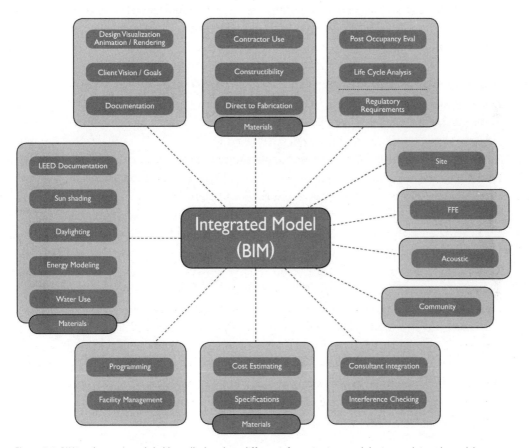

Figure 7.6 BIM involves many stakeholders, all whom have different information input and sharing needs into the model.

If you're not working on an integrated type of delivery method, then you're limited. For example, in a design-bid-build process, the ability to coordinate with the operations team to find out what information they want to see at the closeout of a project rarely happens because the focus is on establishing a budget, addressing constructability issues, and trying to be awarded the project. The unfortunate aspect of a design-bid-build project in regard to project closeout and any record or as-built drawing information is that it will rarely be well-coordinated documentation. Unless the proposal is extremely well written and the facility manager has had an in-depth amount of input into the solicitation, chances are the project closeout documents will continue to be relatively the same quality as what is being seen in the market. Although there are exceptions to this example and some owners and organizations have very well-written project documentation deliverables, the majority of building owners simply don't know what to ask for. Whereas you might have an opportunity to tell owners about the benefits of BIM and in particular the creation of a record BIM at project closeout in a more integrated method, the process of design-bid-build just simply doesn't facilitate that necessary communication between the owner and bidders.

Information in a record BIM is much like a set of instructions. For instance, if you are assembling a bicycle, most instructions will tell you what a piece is, what its function is, and where it goes. A more sophisticated set of instructions will tell you what metal alloy composition the frame is made of and what the recycled content of the tires are.

The needed amount of information varies, because in order to efficiently operate the bike, you need to know how the parts go together but might not necessarily need an in-depth description of every component's chemical composition. However, this might be needed if you were constructing and operating something more complex because the set of instructions would probably be significantly more complex in level of detail and instruction.

Ultimately, what the record BIM needs to contain is any information that is critical to operate it efficiently without spending unnecessary time and effort modeling information that isn't needed. Additionally, current BIM software grows in file size the more 3D information input into the model. Conversely, the fewer model components, the smaller the file size. So, in order to keep your BIM files balanced within a range that a modern-day laptop can handle, it's necessary to identify which items are critical to the design, construction, and operation of a building and which ones can be identified in other ways. One way that information is input into a model without significantly increasing file size is through the simple representation of complex components. This method of modeling creates simple models that have a large amount of information hosted to the elements' properties, or are linked to an external database for reference.

Technology recently redefined element-to-database linking through the use of RFID tags. As mentioned earlier, RFID tags, or radio frequency identification tags, are small identification tags placed on assets that, using a scanner, identify a piece of equipment. RFID technology is more popularly known for the *smart pass* chip that is embedded in credit cards. When a card is scanned, the user's information is directly sent to the credit card company to charge the user for their purchases. Recent developments have seen RFID tags being used in both the construction and facility management arenas. The value that an RFID tag brings to asset management that a bar code does not is that RFID tags, although small, are made out of thick plastic. Bar codes are often stickers that run the risk becoming marred or scratched, leaving them useless. RFID tags can be placed on virtually every asset in a building and can be scanned to pull up all sorts of information about that particular piece of equipment through an editable database. Numerous software programs, such as WiseTrack and IntelliTrack, can be used for asset tracking and provide an open database platform for users to input and customize all sorts of information.

In BIM, the RFID tag uses XML to tie the external databases to the BIM file. Imagine being able to scan an RFID tag, pull up the information about a component, while simultaneously having the software find it in a model. Things such as doors, windows, hardware, HVAC equipment, furniture, and so on, would all be capable of being scanned, and the information could be fed to the user in real time. Currently, there is no software that links to all of the modeling software available, so this is purely speculation at this point. However, it doesn't seem too far of a stretch because companies such as Vela Systems have already begun using these tags to identify construction components and other components in the field.

Other options for linking the BIM to a database include directly inputting information into the elements' properties. This can be done through the use of the software's default fields or the creation of custom fields that can show additional information about a component. Generally, this method is time consuming; however, it removes the need for linking a model component to an external database to drill down to information. In either scenario, the importance is to leverage the BIM to contain the needed information and remain a useable tool for the FM throughout the building's life cycle.

How to Maintain a Model

Just as in the design and construction phases, the value of a BIM file to facility management is in its accuracy and ability to be up-to-date. That said, maintaining a BIM file with regard to facility management information is very similar to maintaining the actual facility. As components are replaced, repaired, or removed, those changes will need to be reflected in the BIM file. The tool needs to be updated to match the facility so the accuracy doesn't fall behind and the BIM file becomes less of a resource for its user. If the BIM file is not updated and relevant, this makes finding the right information the next time more difficult. Without BIM, the facility manager refers to 2D drawings or self-inspects the facility. In the case of larger structures, this inefficiency will be clearly evident and reflected in the upkeep and potentially the stress level of the facility manager.

Let's face it, the BIM file is only as applicable as the user wants it to be. For this reason, BIM and CAD applications rely on the facility manager to update the files over the life cycle of the building. The exceptions to this rule are software updates, outdated data, change of personnel, archiving, and other tasks between the user and the information. Depending on the level of BIM training a facility manager has, the task of updating might require outside help in the form of software vendor support or outside technology consultants. Aside from these costs, however, the data for a facility rests squarely on the facility manager's shoulders. If a facility is to be added on to or expanded frequently, it is critical for the facility manager to keep the drawings and documentation as accurate as possible for future additions and renovations (Figure 7.7). This limits the amount of time and resources that a design team and contractor will need for their due diligence of a facility prior to beginning a remodeling project as mentioned earlier. Basically, the quality of the document directly translates into either savings or expense. And as firms continue to migrate to BIM and as contractors continue to utilize BIM technology, the usefulness of BIM technology during a handoff will continue to increase.

IMAGE COURTESY OF CATES SHEET METAL

Figure 7.7 Example of an existing facility being renovated, which changes the BIM file

Documenting for Decommissioning

Documenting assets for decommissioning is important as the documentation in this phase of a building will be used to quantify everything from scope of demolition to what the demolition contractor can expect to encounter when entering the building, such as electrical lines or plumbing changes. Documentation might also identify hazardous material, grease traps, tanks, or other issues that need special care when being taken apart. The ability to effectively maintain documentation correlates directly to defining the scope of work with the contractor as well. Recently, there has been an increase in civic buildings being decommissioned to meet green building standards. In projects such as these, it is even more important to have the most accurate data on hand because it provides the deconstruction professional with the information needed to deconstruct a building in a sustainable manner. In other words, they need to know what materials can be reclaimed or recycled and what materials must be discarded.

Effectively documenting, operating, maintaining, and completing the life cycle of a facility are the ultimate goal of any facility manager. With a number of other tasks included within these descriptions, there is an enormous opportunity to better define these events. It's pure economics; if a building is designed and operated effectively through documentation, maintenance, and occupant satisfaction, then there is more profit to distribute. The building owner can then effectively use that capital to begin constructing other projects that keep the design and construction industry moving forward. Many designers and contractors fail to realize that the ability for a building owner to develop new projects is directly related to how well a facility performs. Knowing this should directly influence all decisions for the downstream user. The architect should be maximizing efficiency, designing sustainably, and thinking ahead to the future of the building and working with the owner to not only define what the building is but what the building is to become 30, 40 and 50 years down the road. Likewise, the construction manager needs to be thinking about project documentation, operations, and maintenance manuals, as well as communicating and collaborating with the facility manager throughout the construction project to explain the intricacies of the facility before it is handed over. In turn, the facility manager must manage the BIM effectively and use it to operate the facility to its optimum performance while maintaining the health and safety of those occupying or working in the building and maintaining the life safety equipment such as sprinklers, alarms, and so on.

In the end, there are problems and there are enormous opportunities for BIM maintenance in regard to facility management as the BIM process continues to develop. One of these problems stems from the many formats that are used throughout the design and construction process. The architect is using Revit, the contractor is using Navisworks, and the facility manager is using ArchiBus, for example. Whatever the case may be, there currently is limited connectivity between these systems. With the model evolving as it is passed from team member to team member, the end goal is that the product delivered is a refined and accurate set of usable databases. A construction project moves from more detailed design through construction or implementation documents to construction. Often the design and construction teams are unaware of the important role this team member plays in the operation of a facility.

Although interoperability seems like a problem, it is just as much of an opportunity. Within the BIM community, there is a goal to standardize data transfer and develop software that is specifically geared toward BIM and facility management, removed from a CAD-based way of thinking. Although I anticipate that there will be great strides within the coming years in regard to facility management BIM software and computer-aided facility management (CAFM) technology, the process needs to be put in place to deliver complete native-format BIM files as well as all additional information. Additionally, the community should focus on not only how to apply a BIM database to room or spatial validation but also how to use BIM for preventative maintenance, automated systems, component replacement information, and energy analysis, to name a few uses. Although the database of information is important, the facility manager faces the challenge of updating the BIM information with the changes to the facility over the years. One of the biggest challenges in the facility management industry is how to use the data, manage it, and then change the data while maintaining the facility. Current processes don't seem to be working as efficiently as possible, so how can we use new technology to change that? And is BIM the right solution? BIM as a software and as a process will continue to be refined, and facility managers will ultimately benefit from it as a resource. However, there is still work to be done—which starts with the architect and ends with the facility manager—to document and keep the data current for the life of a building.

Investment and Logistics for FM BIM

Over the life of a building, operating a structure correctly means balancing the energy and resources used. The ability for buildings to save energy and resources means that it must operate more efficiently over its occupied life cycle. In the United States, buildings represent 70 percent of electricity consumption (Figure 7.8).

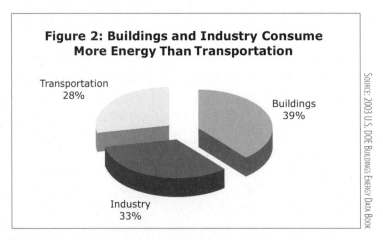

Figure 2: Buildings and Industry Consume More Energy Than Transportation

Transportation 28%

Buildings 39%

Industry 33%

SOURCE: 2003 U.S. DOE BUILDINGS ENERGY DATA BOOK

Figure 7.8 Total percentage of energy use in the United States

BIM is gaining ground and is being more rapidly accepted and required by owners. BIM usage grew by 3 percent in 2003, 6 percent in 2005, and 11 percent in 2006 (according to the Eighth Annual FMI Owner Survey). In 2008, that number grew another 25 percent, and now 45 percent of BIM users say they are using it at moderate levels or higher and are seeing an ROI of 300 to 500 percent (source http://construction .ecnext.com/coms2/summary_0249-296182_ITM_analytics). However, according to the "Cost Analysis of Inadequate Interoperability in the U.S. Capital Facilities Industry" report, contractors and the AEC community at large aren't delivering linked and usable information to downstream users (owners and facility managers):

> *Owners and operators in particular were able to illustrate the challenges of information exchange and management due to their involvement in each phase of the facility life cycle. In summary, they view their interoperability costs during the O&M phase as a failure to manage activities upstream in the design and construction process. Poor communication and maintenance of as-built data, communications failures, inadequate standardization, and inadequate oversight during each life-cycle phase culminate in downstream costs.*
>
> GALLAHER, O'CONNOR, DETTBARN JR., AND GILDAY. "Cost Analysis of Inadequate Interoperability in the U.S. Capital Facilities Industry," NIST 2004

So, how can a construction manager equip a facility manager at the end of a project?

The answer to this question is in the amount of information construction managers compile and, more important, in the manner in which they compile it. It is important to include warranty, cut sheets, and other data about a project, but the game is changing. Costs are rising. As a result, many owners are finding and requesting ways to save and be more competitive through the life of their investment. This will mean that contractors will need to change the information and process of delivering this information so that it can be utilized downstream to stay competitive. The BIM process during the creation of as-built documentation will change. The days of relying on the owner to interpret the mishmash of documents is over. Too many issues are at stake, including cost savings, energy conservation, and life safety, to name a few. The construction company that fully embraces BIM can realize the value of the technology and communicate that to the owner through education.

As discussed earlier, successful document and record BIM delivery involves having the operations personnel actively engaged from the beginning of the project. The operations personnel might very well have information about the Computerized Maintenance Management Systems (CMMS) that they intend to use to manage the facility, and how the documentation and model will need to interface with that software. In addition, the facilities group might have standardized reasons for using the software they do because it might interface with accounting, human resources, purchasing, and other departments within an organization. For this reason, it is important to make sure that the way in which the project is being documented will work with the technology being used in the building.

The need for better model delivery at project closeout is evident in the fact that more than 85 percent of the interoperability costs are incurred during the operations and management phase by the owner (source: www.bfrl.nist.gov/oae/publications/gcrs/04867.pdf). Improved communication continues to be one of the main thrusts behind the industry shift to BIM. However, using too many tools and too many formats with the intention of improving communication can be counterproductive to the group. When looking at a project holistically, take a look at the software that is to be used. Involving the operations personnel early on facilitates better input about the adoption of the model coordination plan and the information exchange plan and should help them define what level of detail and software is needed to perform their tasks.

Training

BIM enables the facility manager to be more efficient; however, there are great strides to be made in the realm of BIM and facility management. With training and some experience, a facility manager can use the record BIM as a tool to more accurately manage a facility than ever before. Because CAD has been predominately the technology taught in universities and technical colleges, it has become the industry norm. For this reason, it is safe to assume that a facility manager who is handed a record BIM who doesn't have any prior training or experience with BIM or its respective pieces of software will have no idea what to do with it.

Therefore, the discussion of training needs to happen early on in the construction process (Figure 7.9), because of a number of reasons:

- First, by the time construction is completed, the facility manager will have a level of understanding to hit the ground running with the record BIM file to keep it current as the facility is handed over.

- Second, the facility manager is apt to give more valuable input into the design and end-level information available in the BIM file and will be able to inform the team what information will be most valuable at the end of a project.

- Last, the facility manager needs to receive the training for the next project. This is ultimately how we all have learned one way or another in BIM—through experience and use. That said, the BIM-trained facility manager is a great resource for document accuracy and upkeep and also for internally creating value for the building owner. The facility manager can now extend his value to other associates without incurring any of the additional training costs for each associate. Although the decision might be to train all staff regardless, if there is a facility manager in place who has been using it, that will be extremely useful down the road when questions arise from the newly trained associates. Implementation planning as discussed earlier in the book is directly applicable to the facility management team, just as to an architecture or engineering firm.

Figure 7.9 Facility managers should be trained during construction so at the end of construction they are familiar with the tools when the project is completed.

Although there is still plenty of room for standardization in BIM—for example, the XML format exchange between all software systems or some version thereof— BIM somewhat negates the old discussions of lines, layers, and plot weights. Of course, the settings can be modified, but the fundamental difference in a standardized format between a CAD standard and a BIM standard in CAD is CAD's line-based process vs. BIM's data-rich object-based process. The task of standardizing is somewhat of an arduous one, especially for an entire industry. Yet, the step that needs to be taken first is the step backward toward defining what BIM's base language is, and then we can move forward—which has been the goal of the NBIMS (National Building Information Modeling Standards) group.

Tower of Babel

Recently I was talking about interoperability with a colleague of mine, Joseph, who is a programmer for Apple. Joseph is a very smart guy, and I revere the fact that he has started up two very successful software companies, sold them, and continues to work for Apple because he likes the "environment." Although Joe isn't intimately familiar with using BIM during construction, he knows more about databases and programming than I ever will.

He had the chance to look at some of the programs we have used in this book, and he commented that "Everything needs to have a base similarity and language for communicating, such as similar data fields and formats, for example. All further revisions must be updated, and new information from different software and functions all must add and refine the information with C++ as the base language and refined file type at the end of a project, or you end up with a Tower of Babel situation in the end."

The discussion stuck with me for a while and I thought it would be interesting to put in the book. While we spoke about other things, he brought up interesting parallels between software companies that were considered "competition"; while they each develop different software, which aim to outsell the other, there is an understanding between them that in order for them to continue to succeed, there needs to be a high degree of interoperability between them.

Implementation and Use

To implement BIM, a facility manager needs to analyze what systems need to be put in place to be effective. For example, if the facility manager is managing a small strip mall, then he might not need more than one license of the native software and probably won't require a sophisticated database to manage the facility. On the other hand, if that facility is a 1200-bed urban hospital, then information systems need to be put in place to allow that facility manager to succeed. In this case, it might require multiple licenses of the native software, a server, sophisticated facility management software, and other infrastructure to support the operations of these facilities. The implementation phase for a facility manager is no different in theory or practice from a construction company, architecture firm, or engineer's office. The general outline is as follows (refer to Chapter 2 for a more in-depth description):

1. Estimate the investment costs.

2. Develop an implementation plan.

3. Choose a manager or enable an existing manager to be BIM lead.

4. Support the user with training.

5. Support the manager with a team.

6. Learn more, contribute to the community, and develop processes.

7. Analyze implementation.

8. Monitor future trends.

Facility managers in general have a thorough understanding and knowledge of database management, and they have in general used CAD drawings as supplemental information or have created their own. Ideally, in a BIM process linking the record BIM file's database to the facility manager's database is where BIM can really create savings and efficiencies. Although this can be done now by using external URL servers and customized software, it requires a significant amount of setup. Future software developments should aim to streamline this process.

In the end, facility managers can be more efficient resources to the building owner and to the industry as a whole. Just as the development of BIM in the construction arena has continued to develop because of dialogue within the profession, introducing BIM into the facility manager arena and promoting discussion within this profession will advance both usership and the tools.

Conclusion

Currently, a record BIM delivery is specific to each construction project. Although there are a number of guidelines for creating a record BIM (such as the GSA's "BIM Guide for Spatial Program Validation" and "BIM Guide Overview" and documentation from the Open Standards Consortium for Real Estate), it has been my experience that the deliverable for each project will be different. This is not necessarily because of a lack of standards or interoperability but rather because the field of facility management and BIM are truly beginning to formulate just what one means to the other. Other project factors such as owner education, staff capabilities, and facility type all have a hand in defining what will be delivered at the end of a project. The keys are to discover what is expected, educate the owner about BIM, and create the record BIM to whatever level is desired and implement strategies throughout the construction process to arrive at this goal.

8

The Future of BIM

This chapter explores the future of BIM in the design and construction industry. Specifically, I will cover the following:

What Will BIM Be?

This chapter takes a speculative look at the future of BIM: where it can go and where it is already headed. It will show what BIM will become in the coming years, as well as theorize on who will use it and how it will change the design and construction community.

To begin, BIM will replace CAD. This isn't necessarily a prediction or speculation. CAD had its heyday, just as the pencil or pen's use in construction documents was once the accepted technology prior to CAD. Our current world moves fast and is connected to seemingly infinite amounts of information that can be accessed in a matter of seconds, and if you don't make the move to new technology, the industry will blow right by you. This simple fact applies to many other industries than the construction industry. Widespread BIM adoption has begun to take hold and sophisticated owners, such as the GSA, see the value of BIM and are demanding it on their projects. As the market continues to adopt BIM as a standard, BIM will continue to flourish. Indeed, what is so promising about BIM is that it is not CAD, it is not drafting, and it is not lines on paper representing the outline of a building. Rather, the model *is* the building.

When Is the Future of BIM?

Why hasn't BIM taken off yet? As an analogy, think of the way doctors performed the first open-heart surgery in 1896. That procedure was quite different from the way heart surgery is practiced now. What changed? What lowered the mortality rates? What made it so commonplace today?

Experience, technology, and process.

Although this book has identified some of the technologies and processes identified with BIM, what will continue to grow is BIM experience—both industry-wide among users and how it is used with new processes to refine the flow of information and responsibilities. As there is a larger base of user experience shared and process based results are discussed, BIM will continue to be refined.

Secondly, composite VDC models will continue to evolve to serve the needs of the team more effectively during design, construction, and beyond. As this book has outlined, current situations require a hybrid use of both 3D views and 2D drafting information to produce construction documents and this will continue for some time. However, as BIM continues to grow in sophistication and experiences are shared, the use and detail of the models will change. There will be no such thing as a *complete* model. Rather, the process of model development will continue to develop during pre-construction, carry into the field as an accurate construction tool and then be used after occupancy in a much more organic and true process tool. The ability to create models that are construction-ready will steadily improve. In fact, there has been recent discussion of why "representative" models are being created at all and if there's any value to them. Currently, the ability to "work" in BIM requires multiple tools and, theoretically, the ability to have all team members continue to build upon previously created models is not often possible. This book has outlined a strategy in the interim that promotes model archiving, updating, and new analysis. Companies like Autodesk are investigating model sharing strategies over the Internet to further develop model sharing strategies. Some companies are developing "BIM servers" and networks that begin to offer single model working strategies. As a result, the ability to build a structure based completely on parametric information will become a reality over the next 5–7 years, just as models and the automation of construction will become a reality for more building components and potentially entire buildings within the next 7–10 years. Figure 8.1 shows a building printed in 3D. Although the technology shown requires only one medium to construct the model, what is truly amazing is the level of detail, accuracy, and sophistication that is capable.

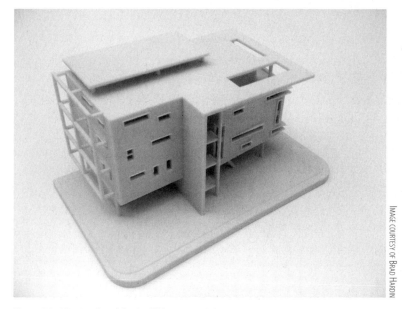

IMAGE COURTESY OF BRAD HARDIN

Figure 8.1 3D printed model using BIM to construct it

It isn't much of a stretch to believe that buildings will continue to involve more and more CNC and automated fabrication. Complex, custom structures requiring a high degree of accuracy as well as simpler structures seeking higher profitability and better coordination will continue to reap the benefits of BIM as a tool for construction and prefabrication. In the future, many construction managers may be responsible for fitting the pieces of a building together as if it were a 3D puzzle on a job site to increase proficiencies and profit. However, the future will still require adept and technology-proficient professionals to carry out this work.

Of course, nothing is perfect, and even if the AE team created the "perfect" model, issues would still arise throughout the construction of a building. The goal is to make these issues fewer. This is why having process *and* teaming agreements in place that facilitate increased communication between team members is critical to the success of the project. BIM is not a liability. My perspective on the industry overall (and in many ways the purpose for writing this book) is that the industry needs to realize that the old processes aren't working in today's world and will most certainly not work in the future. Even if old processes are working to some level, they aren't working as efficiently as they could be. Having seen the results of effectively utilizing BIM from both the architecture and construction sides, it's difficult for me not to become a staunch advocate for using this technology. BIM offers the opportunity for the construction industry to use lean strategies, increase profits, increase coordination, and drive project accountability. Although certain BIM tools could certainly use further development, many companies are currently using BIM and leveraging this technology against existing technologies. BIM is here. Construction companies around the world are using it to sort through complex issues and increase the abilities of a design team. There are no guarantees that using BIM will give a team zero RFIs and the project will be finished without error ahead of time and under budget every time, but BIM will continue to develop. As new processes, delivery methods, and teaming agreements continue to be refined beyond even those mentioned in this book, the opportunities will be incredible.

New Roles

This new process is appealing to a number of stakeholders in the design and construction profession. From the contractor's perspective, there is an opportunity for reduced risk. As a result, this could reduce the contractor "contingencies" for architect and engineering documents with less concern for change order exposure. Using the build as modeled philosophy, the contractor will have a much steadier foothold on otherwise shaky ground. On the other hand, the architect has more responsibility to create a more detailed and accurate model. There are many perspectives on the increased responsibility and perceived liability associated with the architect and contractor collaboratively creating a construction-ready model; here are two main ideas:

- The first perspective is that design professionals should continue by using the same processes as they always have and let the liability of document interpretation fall on the contractor's shoulders. Many of the statistics in this book show this process lacks document quality and increases cost.

- The second perspective is that there is a serious risk in the profession of architecture in continually forcing more responsibility on the contractor by not accepting any and, as a result, more profit. Some even believe that the professions of architecture and engineering might face some level of extinction as a result of the industry unwillingness to take ownership of their designs through increased responsibility, instead leaving it up to subcontractors, fabricators, and contractors to virtually build their designs. While this is a large assumption, this concern has some merit evident in the steep rise in the hiring of architects and engineers to become part of virtual construction teams working for general contractors. Not only have more BIM professionals been hired by general contractors, but specialty trades such as sheet metal, mechanical, electrical, piping, and structural steel fabrication shops are employing BIM specialists as well to meet the rising BIM demands. Additionally, the popularity of one-stop shop entities (such as Jacobs Associates, Parsons Brinckerhoff, and CH2M HILL) that offer in-house design and construction BIM services makes sense to many owners as the process becomes streamlined and the responsibility for the project is a team initiative, which further blurs the lines between professions.

Regardless of the new role of the architect, engineer, and contractor, the industry has begun to accept that new teams and responsibilities need to be created in order to realize the full potential of BIM. Also, there are now more possibilities which influence how a project's fee is structured. Environmental analysis, code testing, and material use can be reviewed more quickly than before. Although these services can make for a better project, they should be carefully considered and outlined, as a team can become inundated with analysis and lose efficiencies on a project. On the other hand, teams can also be reimbursed for such services, which ultimately bring value to the owner and the project as a whole.

In the future, more teams will become aware of the wide range of benefits that come with partnerships. BIM has already changed the face of the construction industry. There wasn't as much talk of enhanced capabilities, reduction of errors, and future advancements with CAD with nearly as much vigor as we are seeing with BIM. Simply put, the enthusiasm has now been usurped by the reality and details of making a BIM process work. While this is in no way an overnight effort, huge strides are being made in both advancements in software as well as process changes and new concepts of delivery and teaming that strategically alter the way we design and construct buildings.

Interoperability

BIM will play better in the coming years. Current interoperability issues require many pieces of software and many workarounds that take additional time and might or might not be able to be updated. This will continue to change, particularly as the International Alliance for Interoperability continues its work and as software companies become interested in making their proprietary product more effective than the competition.

The future of our industry will continue toward a more integrated course, yet the outcome of this integration is still to be determined. Some believe that general contractors will continue to merge and acquire design firms with the intention of streamlining practices and increasing profits. Others believe that project alliances will continue to gain market share and increase in popularity among the design and construction industries. The bottom line is that our industry will continue to change.

As a result of this change, systems, ERP workflows, and transfer rates will need to work between programs. Organizations such as the buildingSMART and International Alliance for Interoperability (IAI) champion these critical efforts. In the future we will continue to see more success stories such as the IFC file formats. There is an enormous potential for synergistic relationships to exist between the working construction industry to be represented by organizations working directly with companies creating the tools. For example, CSI's work on the Omniclass classification system, which provides a new structure to electronic databases, provides software designers a basis to construct the framework of their software. CSI has committed to developing the International Framework for Dictionaries (IFD) Library, that provides "a real opportunity to generate an IFC-BIM for operational and maintenance purposes with storage of product specific data, a feasible method of linking existing knowledge systems to an IFC-BIM, and multilingual and translation capabilities to the information in an IFC-BIM." Additionally, the buildingSMART alliance is doing important and critical work in developing and defining the National BIM Standard (NBIMS) and has created a forum in which users are sharing their experiences and communicating results and recommendations.

Interoperability in the next 5–7 years will see great strides as the base of how BIM communicates between systems will continue to be integrated and developed. Although it's difficult to speculate as technology moves so fast, in 10 years we should see a reduction in the relative number of software tools available as they will be integrated into existing software or available on an as needed basis via the Internet. Additionally, the disparity between how systems work together will be significantly reduced, as well as the costs associated with the purchase of multiple pieces of software to accomplish some of the tasks associated with construction management.

BIM and the Future from the Past

Imhotep first drew lines of ink on papyrus to indicate the outline of a structure, because this was the technology available to him (Figure 8.2). If the technology available to him had been a better technological advancement, would he have used it? Probably. He used ink because of necessity and availability. He might have asked himself, how can I communicate this design in my head to the workers who are building it? As this may have been a question of logistics, the question the modern construction industry is asking is, how can we *best* communicate the design and construction of a structure to everyone?

Figure 8.2 Imhotep, the world's first recorded architect

Many people question how this new technology is better than the "good ol' days." Although I respect those who have accomplished so much in our industry in the past, there is clearly a need for change.

For some time, I had heard that the aerospace and automotive industries have been using a form of BIM, rather—parametric modeling. I decided to research what was actually being accomplished; I thought that if the aerospace industry has been doing it for more than 11 years or so, then that industry must have reached some sort of integration from which the construction industry could draw parallels. I received some mixed results. To begin, the parametric software that aerospace engineers use to model is similar to how the construction industry is using BIM software currently. Some pieces of software are used to model and create, and other software is used to test and analyze. What is fascinating about the evolution of the way engineers have been using modeling and CAM technology is that they found it was virtually impossible to create an aircraft that contains 4 million parts while working from remote locations. In addition, I was amazed at how similar the process of designing aircraft was to the process of constructing buildings.

According to "Aircraft Design: Synthesis and Analysis" from Stanford University (http://adg.stanford.edu/aa241/AircraftDesign.html), the purposes of the computational model are as follows:

- Allow the simulation of the behavior complex systems beyond the reach of analytic theory.

- Provide detailed design information in a timely fashion.

- Enhance our understanding of engineering systems by expanding our ability to predict their behavior.

- Provide the ability to perform multidisciplinary design optimization.

- Increase competition, and lower design/production costs.

Sound familiar? The results of what we are trying to accomplish with a BIM file are almost direct parallels to what the aircraft industry is trying to accomplish with its 3D parametric models (source: www.infoloom.com/gcaconfs/WEB/chicago98/baum.HTM). After digging further, I came up with the following lessons learned from the aircraft industry. According to the report, they are as follows:

A solution is only as good as the model that is being solved. If you try to solve a problem with high nonlinear content using a computational method designed for linear problems, your results will make no sense.

The accuracy of a numerical solution depends heavily on the sophistication of the procedure employed and the size of the mesh (model) used. Lower-order methods with underresolved meshes provide solutions where the margin of error is quite large.

The range of validity of the results of a given calculation depends on the model that is at the heart of the procedure. If you are using a solution procedure to approximate the behavior of attached flow but the actual flow is separated, your results will make no sense.

Avoid information overload. Computational procedures flood the designer with a wealth of information that sometimes is complete nonsense! When analyzing the results provided by a computational method, do not concentrate on how beautiful the color pictures are. Instead, be sure to apply your knowledge of basic principles, and make sure that the computational results follow the expected trends.

In essence, the problems the aircraft industry has faced with parametric modeling correlate to the construction industry. The previous list can be summarized as follows:

- The information you are using is only as good as the virtual model constructed.
- The higher the level of detail, the more accurate the construction. The more generic the model, the more room for errors.
- If you're not using a model with the same base language, teamed with the right processes, the fluidity of the process will be disparate, and this will be reflected in the model.
- The amount of information in the model needs to meet minimum standards, usable by everyone in the team. Conversely, creating loads of useless information doesn't help anyone, nor file sizes.

Taking these lessons learned from the aerospace industry, we can chart a path for BIM and its future in our industry. The use of BIM is unique in construction, of course, because we work on one project with a team and might not work with that team ever again—unlike prototypical production such as an airline manufacturer. However, the simple truths of integration, interoperability, better communication, and new processes are the common threads for both industries.

Sometimes in the construction industry we lose perspective and believe that we have the most complicated and sophisticated profession in the world. Although there are many challenges in engineering, phasing, design, and construction, we need to remember that our industry involves people working with people. Better projects are a direct result of people working with better processes enabled by better technologies. BIM is only part of the answer and will continue to be just a portion of the solution. There seems to be a belief that BIM will solve all of the ails of the construction industry. Although BIM will help considerably, it's up to us, a group of construction professionals, to enable BIM and other future technologies to help us work better together by learning from each other and the past.

BIM and the New Construction Manager

BIM's biggest opportunity in the future will be the direction that new personnel take BIM technology. Sophisticated owners are seeking out design and construction companies that utilize BIM to meet their needs. Because educated owners will continue to award projects based on qualifications, team selection will continue to be critical to the success of these companies. These BIM enabled companies rely on the professionals they hire who make up these teams to deliver results. Choosing the right associates are not only critical in beginning a BIM process, but these team members will be the ones to refine it over time. Wanting everything to be BIM enabled from the word *go* is not a reality and takes time to develop.

Our industry needs to continue to determine how BIM technology can increase profitability by finding associates who can produce process-based results using these various BIM technologies. The need for measurable results does not necessarily need to be compared with CAD anymore; rather these associates need to begin comparing BIM enabled projects with other BIM enabled projects. The case for BIM has been made. And when direct comparisons are made we will begin to see results that compare apples to apples in both process and technology refinement specific to BIM. When direct analysis is accomplished, we'll see a much clearer picture of where we can currently be effective and where the technology needs to grow and in which direction it should grow. With the focus on developing and defining procedures, we'll get the power of people sharing information and experiences and lessons learned along a related timeline and defined milestones. In addition, entrepreneurs will begin to develop the software, resources, and dialogue necessary to meet the industry's needs that are defined along the way. This is critical to the success of BIM in our industry and is a driving force behind writing this book in this exciting time of education, research, and implementation of this new technology.

So, who are these people capable of utilizing BIM technology, and where will they come from? Just as important, who currently is filling this industry need? Currently the industry is hiring architects, engineers, and construction professionals who have received specialized training, or have prior experience in BIM. This new industry need has created an extremely competitive market, as the availability of professionals with BIM experience is limited. As a result, many companies have sought out other resources and BIM consultants to engage in their projects. These consultants are often highly trained and effective, with varying degrees of costs. Consultants rarely offer training, as it is their business to bring the expertise to the table, which could limit their revenues if firms knew what they know. This gap in the industry has brought the focus to newly trained professionals. Although being trained on a number of BIM programs in order to be effective is a matter of time and experience, future generations are entering the workplace with skill sets that are needed in the industry.

Future generations of people who use more advanced tools will be the ones who continue to develop BIM. These future architects, engineers, and field and construction managers will begin to cross conventional industry lines and focus on successful project delivery. In fact, it is in their very nature to leverage technology to access information and work together as teams to accomplish tasks (Junco and Mastrodicasa, "The .NET Generation"). This future generation expects ease of access to information: over 97% of them own computers and over half of them use blogs to disseminate information among their peers and colleagues. In the construction industry these new professionals are part of a larger interdependent relationship, in which they will require mentors to guide their intuitive and proficient use of the software tools by providing insight using experiences and lessons. Additionally, these professionals will bring a level of understanding and comfort with software and technology that more senior members of the team might not have. Moreover, they will be focused on utilizing all resources available to deliver projects. In this book, I have talked about the virtual construction manager, or BIM manager. The rapid expansion in this previously nonexistent field proves a further testament to the need for these associates and their abilities to leverage technology to help a team better coordinate projects. And there will continue to be shortages of these types of individuals until schools begin to adopt curriculum that teaches students about BIM and its tools, as well as informs them of current-day realities and needs in the construction industry.

Education

For many, BIM training begins in academia. Education, particularly in universities, is where the ability to create new mind-sets and exposure to new media is most effective. Schools teaching architecture, engineering, and construction management have an enormous responsibility and challenge to present their student body with technologies and processes that are relevant to the environment they will be exposed to upon graduation. These institutions must realize that they are just as important as the company the young graduate will work for after leaving the institution:

- First, a college or institution is often noted for what level of technological advances the school has embraced. If the university is deaf to the industry and its trends, this will be carried through in the school's reputation as well as its graduates. For this reason, it is critical that university systems are grounded in current issues and technologies, because these will be the issues their graduates will face in the industry.

- Second, institutions have the ability to be extremely advanced. There is little to no risk in exposing students to new technologies. Exploring new software and different processes should be done at this level, because it pushes the students to learn and formulate views and solutions of their own. This ultimately will enable them to hit the ground running after graduation. This education might not result in a complete understanding of all BIM tools in the architecture or construction profession, but it should include a base level knowledge of what the software is capable of doing and how it is to be used.

- Lastly, universities have the potential to save companies training costs. Although this shouldn't necessarily be a goal for universities as there are so many different pieces of BIM software, it is a marketable byproduct of embracing and training its student on new software and technologies.

Fundamentally, BIM thrives in a collaborative environment and when utilized early in the design process. In the future, schools will need to create an environment of collaboration, which will better prepare its graduates for what to expect outside of its doors. This will enable students to better understand the actual working methodologies of construction teams and learn that very few professionals work on their own. To some extent schools must test their students individually; however, creating team-focused groups with goals for success will prepare the new professional for hands-on practice as well as the opportunity to explore groups that cross disciplines. Universities have a unique opportunity to simulate real work in that many schools have architects, engineers, and construction students on hand that bring a differing perspective as well as a differing set of goals to the learning process. Understanding the importance each player brings to the team lays a foundation of collaboration as well as a frame of reference as to what to expect in the industry after graduation.

BIM software is complex. And understanding the ins and outs of a particular tool often takes years. With this in mind, it is important in education to expose students to a basic level of understanding about how these tools work and what they can be used for. Although some students might continue to learn about other pieces of software after graduation, it's important for schools to teach BIM at a conceptual level. Often students will begin learning whichever software is chosen to be part of the curriculum and will self-teach to a more advanced level. This is a unique characteristic of the new generation X and Y'ers. This generation has the ability to learn software by simply working in it, pushing icons and working within the interface which seemingly comes naturally. As the generation has achieved a level of comfort with technology, such as the Internet, video games, and leveraging the power of computers for advanced problem solving, this is sure to continue as a trend for generations to come. In essence, instructors cannot afford to believe that BIM tools are too complex and varied, and avoid spending time researching and understanding the software. It will be critical for instructors and schools alike to embrace some level of BIM technology and introduce students to some of these programs and concepts, which can add a new layer of knowledge to a student's learning and increase their overall marketability.

Schools should also investigate the industry holistically. Although there are areas of focus such as design for architects, math and calculations for engineers, and management classes for construction management students, many other job opportunities now exist in design and construction. Architects now are being asked to use BIM to design sustainable buildings that capture rainwater, use less energy, and are more eco-friendly over the life cycle of the building. Engineers are being asked to utilize BIM to calculate complex CFD equations, find what type of steel is most efficient and least wasteful, and analyze information from a building to measure performance. Contractors are switching focus from just construction management to life cycle management and are leveraging BIM to maintain accurate information about facilities that require frequent expansion and renovation for their clients. These are all real-world components of the modern AEC industry, and schools have a great opportunity to begin passing this information on to students.

Workplace

BIM is an enabler of on-the-job education. By having the ability to understand how buildings are constructed using 3D technology (Figure 8.3), professionals just starting out are put on a fast track to learning through tools, which help them understand actual construction (Figure 8.4). In addition, BIM offers an interdependent relationship between the senior staff and the new staff.

The introduction of BIM technology has brought about new trends in the workplace. One trend is the interdependent relationship that is formed from the adoption of BIM between newer associates and senior associates. The senior associate brings experience, both real-world and practice-related, to a project, and the younger staff brings a comfort level with technology and the ability to perform tasks and compile and manage data. The younger staff is dependent on the direction of the senior staff, and the senior staff is dependent on the younger staff member's ability to virtually construct, analyze, and provide results. This improved methodology of apprenticeship could potentially train new professionals quicker and more effectively, all the while using BIM technology to help the senior associate to make better-informed decisions. Just as BIM is utilized during the construction of a project, it can also be used to improve communication and collaboration in education and the workplace as well.

Figure 8.3 Cerner clinic design

Figure 8.4 Cerner clinic, completed construction

Continued Definition

Throughout the book you have heard the terms *BIM manager/director*, *virtual construction manager*, *VDC coordinator,* and so on. In many ways, this new position is responsible for virtually constructing, analyzing, and managing a model. The responsibilities assigned to this role vary from each company, yet the following is somewhat of a recurring theme in the job requirements for a BIM manager:

- Take responsibility for the entire scope and quality of the BIM team.
- Ensure adequate personnel is available for each BIM project's needs.
- Recruit qualified modeling personnel, and provide management assistance to the BIM recruiting efforts.
- Participate in selecting BIM projects to be pursued.
- Attend preproposal meetings, perform site visits, and collect data required to perform modeling tasks.
- Assist with responses for requests for proposals by being familiar with industry standards and terminology as well as listing past project experience and examples.
- Prepare conceptual and detailed modeling budgets for proposed or awarded projects, and review them with the operations manager.
- Monitor and manage BIM expenses, and review them with senior management.
- Assume key roles in select sales presentations for assigned projects.
- Identify the modeling team for active or proposed projects.
- Ensure the responsibilities of others and assignments are coordinated and met.
- Guide model creation/assimilation throughout the preconstruction process.
- Provide assistance to the estimating team for model review and RFI model documentation.
- Take the lead role in facilitating, conducting, and participating in project kick-off meetings, design meetings, preconstruction meetings, and the operation's project meetings.
- Establish credibility and confidence for BIM as needed with clients, design team, subcontractors, and internal team members.
- Maintain and expand various design relationships and partnerships.
- Develop and ensure R&M BIM policies and procedures are implemented and followed on projects.
- Provide guidance in regard to design management and model completion.
- Conduct detailed model reviews to ensure project expectations are met.
- Ensure the preparation of clear, properly formatted model quantities and material takeoffs.

- Ensure questions regarding projects and model information are properly documented and addressed to the design team.

- Produce conceptual and programmatic models based on similar projects and historical cost information.

- Ensure measures are in place to monitor and track changes, and provide information to internal team members, design team, and subcontractors.

- Proactively identify risk factors with senior management.

- Assist operations with the development of construction phasing models throughout the project as necessary.

- Assist operations with the development of project site utilization modeling and logistic coordination.

- Encourage the exploration of innovative, technically creative model presentations as needed.

- Strive to attain 100 percent customer satisfaction for BIM projects.

My Job Changes Every Day

"Every day I step into the office, there is something else I'm going to learn, the team is going to learn, or the industry is going to learn. The fact that we continue to find ways to leverage BIM technology during the design and construction process means that we still are realizing its potential and have yet to find all of its applications. This is both an exciting time and a time to understand that this is when most of the work needs to be done so that future generations of design and construction professionals aren't realizing the same issues with interoperability, cost, and applications that we are."

—BRIAN ARNOLD, BIM specialist, McCownGordon Construction

Whether the title of the BIM manager becomes *information director* in the future or whether it goes in another direction, within the framework of BIM the BIM manager will have the opportunity to become a unique entity who consults with architects, contractors, and facility managers. Additionally, BIM teams will continue to be internalized and used to produce efficiencies, and they will continue to become a critical piece of the overall practice and marketing strategy of a company. The roles and descriptions of BIM users will continue to evolve, just as the software does. To this end, it will be critical for companies to be constantly aware of new software, trends, and delivery methods, as well as the best personnel to support the organization. Technology is unique to the construction industry in that there are so many unknowns and variables for each project, and an industry-wide process of delivery and single software solution seems like a long shot. Yet this industry will continue to become more standardized in the use of BIM, and the use of new technologies teamed with new processes will continue to advance the profession.

BIM and the New Team

BIM encourages the formation of new teams in the industry. In the form of more integrated, all-in-one shops and the creation of individual project-driven teams, BIM will continue to drive the industry to refine the way it works. Building information modeling as a technology has the ability to transform our industry and the future of other industries, such as GIS, accounting, project management, emergency response, GPS, and environmental applications. The fact that there are more and more intelligent models created means there is more information. When this additional information is more effectively shared and linked to other systems, additional savings, and reduction in duplicative work, will be realized. Currently, BIM is beginning to focus on the construction industry, but as it grows it isn't much of a stretch to envision the following scenarios:

- Firefighters pulling up the building's 3D model from a city's database as they are driving toward the fire could be able to identify exactly where all the fire extinguishers, shutoff valves, and emergency exits are in the building before ever stepping off the truck.

- Code review professionals could perform more accurate code reviews through the use of software (such as Solibri) to identify necessary clearances, heights, ratings, and appurtenances.

- Government agencies could map their GIS systems with 3D information to show zoning, show adjacent heights, and use and simulate natural disaster management, attack, and energy failure, among others.

- Environmental agencies could simulate buildings' energy use, heat gain, and carbon footprint.

- Nonpublic information, such as government organizations and armed forces, could use BIM, RFID and 3D mapping to manage assets and personnel equipment and to simulate everything from temporary base setup to operation simulation.

Ultimately, a BIM is a usable tool for many other stakeholders involved beyond the design and construction communities. BIM will continue to become a better resource than previous technologies for opportunities to mock up, fabricate systems, link to other software, and remove unknowns prior to physical construction (see Figure 8.5). As BIM is still being defined in many ways in the construction industry, over the next decade we can expect to see other groups find purposes for which BIM can be used to benefit them as a result of a single current database of information.

Figure 8.5 Ronald McDonald House built using BIM and prefabricated mockups

Increased data flows will create an entire workforce dedicated to data management. This enhanced data control will come in the form of virtual design and construction teams. As newer staff members continue to rise within organizations around the world, the comfort of working with such technologies and the means of increasing the individual's effectiveness will create much leaner and more focused groups. The value of professionals using software to construct, test, simulate, and analyze prior to erection will continue to increase and shift the norm. Likewise, the tools will continue to increase in sophistication and usability as more tools become interoperable and as the industry defines its processes and needs. The new team will have the benefit of industry BIM experience and will be responsible for collectively delivering record models to the owner either as a joint venture between the architect and contractor or as a separate but connected entity providing support to the owner, architect, or contractor.

In many ways, this has already happened within the design and construction communities. Design-build firms currently leverage their ability to deliver projects faster and more efficiently because all of their operations are under one roof. Although it is pure speculation that data transfer rates will increase, the future of the BIM team will probably be one that can collaborate in real time and in the same virtual space. Recent developments, such as Autodesk's attention to an Internet-based model approach, as well as "BIM servers" capable of streamlining the task of working on a single model, show the direction of the industry's focus on BIM. Currently, modeling and designing in real time from different locations is not a reality. Trying to do so requires the implementation of strategies outlined in this book to update and archive models and to limit rework between team members. In the future, technologies will be developed to eliminate the current shortcomings in BIM. More importantly, there will be new teams with new perspectives to further enable BIM technology developments.

BIM and the New Process

Throughout this book I have talked about process. Creating a BIM process depends on the current software available, but at the same time changes made to the design and construction of a project must be profitable in order to be worthwhile. Process change, technology implementation, and the addition of personnel resources all take time to develop. Shifting from CAD technology and CAD thinking is something that will be done over time, not overnight, and there are many hurdles to getting BIM integrated into a company. Implementation strategies can easily take years to fully realize, staff is difficult to find, and using BIM requires the use of new processes. Additionally, new technology and tools continue to enter the market and might further shift processes and thinking, perhaps even beyond BIM as we know it today. However, when compared to the potential results, these challenges make the case for BIM. At times, BIM can be daunting and somewhat overwhelming due to the new influx of information and tools. This does not mean that there is room to be laggard about beginning down the path to implementation and still remain competitive in today's construction market. The reality is that BIM is more efficient than CAD. It is a more efficient way of documenting, modeling, testing, analyzing, estimating, and constructing. BIM will continue to gain market share and its users will continue to be selected by educated owners, fellow professionals, and supported by facility managers. Process will continue to be more clearly defined.

When you define process in tandem with available software systems and milestone tasks, you create a road map for future developments as well as a tool to learn from. BIM is not the end-all solution to technological developments in the design and construction industries. If BIM is to succeed, it must continue to be a conduit of change by transforming the way we practice design and construction. In this regard, BIM has been an invaluable tool in opening the dialogue about new delivery methods and collaboration methods and in questioning our existing processes as an industry to find a better way to deliver our services.

Future Opportunities

BIM continues to grow with new applications and new abilities. The past decade has seen such a rise in BIM applications, and although it might be hard to believe that more tools will enter the market, the current tools are just a small portion of those that we will see over the next two decades. For example, some of the tools that exist today are automated code-checking, clash detection, simulation, estimating, and CFD mechanical software—all developed within a relatively short time frame. For this reason, it isn't hard to anticipate within the next 10 years we will begin to see huge strides in the development of BIM software and applications, with process, experience, and market demand directing the industry. The next 20 years could see a completely digital means of delivering a construction project that utilizes no paper drawings and solely relies on a well-constructed BIM file for everything from "e-permitting" with local municipalities to contractors directly ordering materials from downloaded BIM components.

There are huge opportunities within the manufacturing market to begin creating parametric components that contain embedded specification information and are dimensionally accurate. These components should automatically host to other components based on the applicable code. Future possibilities might be fire-stopping tests that verify the UL rating of a downloaded assembly in a two-hour wall and report whether there are any discrepancies, such as one-hour doors hosted in three-hour walls that generate an error, and windows with embedded specification information that does not meet the required natural light levels and sends a notice to the designer. This sort of BIM automation will create a new level of design and accuracy that has not yet been achieved. Although this will require an investment on the behalf of manufacturers, many manufacturers will undoubtedly see this as a temporary opportunity to gain an edge over their competition (until they create their own). Additionally, manufacturers will be able to limit the amount of paper used for brochure and catalog printing, establish themselves and their commitment to the design community, and provide a resource that further promotes their products. Potentially, third-party websites that link to all of these manufacturers may begin to rise in popularity as well.

In the future, we will see more software combining multiple analyses and testing tools into a single product. This will in turn expedite the efforts being put forth by the design and construction teams. Multiple analysis test beds (MATs) that can test for constructability, model integrity, and clash detection at the same time will be needed, in part because of the availability of other tools entering the market and the need to streamline the process of testing the BIM thoroughly prior to construction. MATs will be able to test a model simultaneously for daylight and heat gain, system performance during peak needs, and CFD of the building design. Because of the number and cost of BIM tools, combining these resources into a single platform will be somewhat of a necessity. In the future we will be able to send a model via the Web to a series of linked MATs and return after the model has been tested by a large array of tools. While companies like Green Building Studio offer this service, many other companies and testing software will follow suit and adopt an online test bed strategy which design teams can plug into. Ultimately, this will better enable the designer and builder to make better decisions on many fronts.

Future Relationships

In his book *How Buildings Learn*, Stewart Brand describes how buildings evolve and are refined and altered over time, arguably creating better architecture and purpose. What's interesting is how a design that was originally intended for a singular use evolves. For example, the factory became a warehouse that then became artists' lofts, which then became condominiums and first floor storefronts. Structures can be quite complex, ranging from hospitals to factories to laboratories, and change over time. Additionally, owner's needs, construction materials, construction technology, and available resources are constantly changing as well. Constructing and remodeling facilities requires an advanced understanding of codes, construction methods, safety, materials, design, use, construction scheduling, and holistic lifecycle thinking. Although it is possible, it is highly improbable that an architect will have the required construction and field experience to successfully construct a building and coordinate equipment, subcontractors, vendors, inspections, and safety protocols, all while maintaining the project schedule. It is just as unlikely that a contractor will have an understanding of codes, design logic, programming, accessibility issues, material properties, fire assemblies, and so on, to successfully design and document a facility prior to construction. Thus, we will continue to need collaboration now and in the future of our industry. However, in the future we will see an increased need for flexible teams who continue to blur the lines of common practice and push the boundaries of integration and use of technology.

By combining the knowledge base of the contractor with the knowledge base of the architect and engineer, you can begin to create a recipe for successful project delivery. BIM is a platform that enables collaboration and allows both groups to learn from each other as the project is being developed and to be more quickly informed. Both contribute their information to the model, and using technology to coordinate and remove some of the human error, they can form a project-focused team. In this new relationship, documentation changes and communication is paramount. Similar to medieval times, when master masons used physical scale models to convey to the builders what was to be constructed, modern construction professionals can now use a virtual model to communicate to the subcontractors what is to be built and how.

The future of BIM is both exciting and challenging. Current opportunities are allowing us to bridge the gaps in a disconnected process more and more. This begins in new education standards and carries through to new ways of managing projects, mentoring, marketing, ordering, managing and quantifying results. It has been my experience in working with BIM that there are many more opportunities than shortcomings, and many more possibilities than liabilities. In construction, very little is constant and things are always changing. Ultimately, the flexibility this industry requires needs to be reflected in the technology and processes utilized by its professionals. Additionally, the professionals in the industry are critical to its success. The education and training available to them and the experiences gained and shared are critical to the growth of a BIM enabled process. Industry professionals who have a working knowledge of the needed tools and problems to be solved will continue to dialogue about the necessary gaps and new methods which should continue to develop the software required. Building information modeling has proved to be an extremely valuable tool to many in construction and will continue to be one of the most exciting developments in the years to come.

Appendix:
Applications Supported by Navisworks

The following tables, one per software company, show programs and file formats that are compatible with Navisworks.

In the table, column headings mean the following:

Export A check in this column means there is an export filter available for the software that converts the native model into a Navisworks model. This tool may be an optional plug-in that adds a menu item.

Native A check in this column means object properties of the native model are supported in Navisworks. This may include items such as material colors, layers, and other specifics.

Properties A check in this column means Navisworks can import the properties associated with an object and insert them in the properties tab in Navisworks.

Other An entry in this column means the file must be exported to one of the listed file formats so that it can be imported into Navisworks.

Content of this appendix is courtesy of Autodesk, used by permission.

Product	Exporter	Native	Properties	Other
AutoCAD R14 to 2009 Models can be exported from AutoCAD as either .nwc or .nwd format, depending on whether you have a JetStream Roamer, or JetStream Roamer + JetStream Publisher license respectively. JetStream also supports the native AutoCAD .dwg file format, which can be read directly into JetStream. Object properties, (such as Material Colors and Entity Handles) are also supported by JetStream and these can be viewed in the 'Properties' window. To open the Properties window, go to View > Control Bars > Properties.	✔	✔	✔	DXF 3DS
ADT 3.3 to 2008 In order to read an ADT file into JetStream, you either need ADT or the ADT object enabler installed on the same machine. NOTE: If you have ADT 2007 installed on the same machine as JetStream v5, you will also need to install the ADT 2007 Object Enabler. This relates to all AutoCAD 2007-based products. If an earlier version of ADT is installed (pre-2007), the object enabler is not necessary. From ADT you can export to .nwc or .nwd depending on whether you have a JetStream Roamer, or a JetStream Roamer + JetStream Publisher license respectively. JetStream supports ADT object properties and these can be viewed in the 'Properties' window. To open the Properties window, go to View > Control Bars > Properties. NOTE: JetStream supports ADT textures, though you will require a JetStream Roamer + JetStream Presenter license.	✔	✔	✔	DXF 3DS
Building Systems In order to read a Building Systems file into JetStream, you either need Building Systems installed on the same machine as JetStream, or AutoCAD installed complete with the Building Systems object enabler. NOTE: If you have ABS 2007 installed on the same machine as JetStream v5, you will need to install the ADT 2007 Object Enabler and the ABS 2007 Object Enabler. This relates to all AutoCAD 2007-based products. If an earlier version of Building Systems is installed (pre-2007), you do not need to install the object enabler. JetStream can export to .nwc or .nwd depending on whether you have a JetStream Roamer or a JetStream Roamer + JetStream Publisher license respectively. JetStream supports Building Systems object properties and these can be viewed in the 'Properties' window. To open the Properties window, go to View > Control Bars > Properties.	✔	✔	✔	DXF 3DS
Inventor 5 to 11 In order to read an Inventor file into JetStream, you either need Inventor 5 (or higher), or Inventor Design Tracking 5 (or higher) installed on the same machine as JetStream. Inventor Design Tracking can be downloaded from support.autodesk.com. If you are using an Inventor Project file, (*.ipj) you will need to set this as the Active Project in JetStream, prior to opening your assembly drawings (*.iam). You can set the Active Project either by opening your *.ipj file in JetStream, or by going to Tools > Global Options > Inventor, then enter the full path and file name of your *.ipj file (e.g. C:\my Inventor projects\project 1.ipj). Once loaded, you can then Open the corresponding assembly drawing(s) as usual.		✔		IGES STEP

Continues

Product	Exporter	Native	Properties	Other
MDT 6 to 2008	✔		✔	DXF 3DS VRML
The MDT native .dwg file can be read directly into JetStream, however you will need to install the MDT Object Enabler (available to download from the Autodesk website).				
You can export a file from MDT to .nwc or .nwd depending on whether you have a JetStream Roamer, or a JetStream Roamer + Publisher license respectively.				
Object properties, (such as Material Colors and Entity Handles) are also supported by JetStream and these can be viewed in the 'Properties' window. To open the Properties window, go to View > Control Bars > Properties.				
Revit 4 to 7			✔	DWG DXF DGN
Revit files can be viewed in JetStream if they have been saved as either a .dwg, .dxf, or .dgn file format.				
If you use the .dwg Export out of Revit 6 and 7, then object properties such as Element ID, Type and Category are supported.				
Revit Building 8.1 to 9.1- Revit Architecture 08-09	✔		✔	DWG DXF DGN
JetStream can export .nwc files directly from Revit Building.				
As well as having Revit Building and JetStream on the same machine, you will additionally need to install the .NET Framework. The installer for the relevant version can be found via Menubox (on the JetStream v5 product CD and downloadable installer).				
Revit properties are supported. These can be viewed in the 'Properties' window. To open the Properties window, go to View > Control Bars > Properties.				
Revit Structure 2 to 4	✔		✔	DWG DXF DGN
JetStream can export .nwc files directly from Revit Structures. As well as having Revit Structure and JetStream on the same machine, you will additionally need to install the .NET Framework. The installer for the relevant version can be found via Menubox (on the JetStream v5 product CD and downloadable installer).				
Revit properties are supported. These can be viewed in the 'Properties' window. To open the Properties window, go to View > Control Bars > Properties.				
Revit Systems	✔		✔	DWG DXF DGN
JetStream can export .nwc files directly from Revit Systems. As well as having Revit Systems and JetStream on the same machine, you will additionally need to install the .NET Framework. The installer for the relevant version can be found via Menubox (on the JetStream v5 product CD and downloadable installer).				
Revit properties are supported. These can be viewed in the 'Properties' window. To open the Properties window, go to View > Control Bars > Properties.				
3ds Max 3 to 9	✔			DWG DXF 3DS IGES
To view your 3ds Max model in JetStream, you will need to have 3ds Max installed on the same machine. Then, simply export to .nwc for use in JetStream.				
JetStream does not support .max files, so you cannot read the native file format directly into JetStream.				
NOTE: In order for 3ds Max materials and textures to be exported, you will require a JetStream Roamer + JetStream Presenter license.				

Continues

Product	Exporter	Native	Properties	Other
VIZ 3 to 2007 To view your VIZ model in JetStream, you will need to have VIZ installed on the same machine. Then, simply export to .nwc for use in JetStream. JetStream does not support .max files, so you cannot read the native file format directly into JetStream. NOTE: To Export VIZ materials and textures you will need a JetStream Roamer + JetStream Presenter license.	✔			DWG DXF 3DS
Maya JetStream has file readers for .dxf, igs, .iges, .wrl, and .wrz.				DXF IGES VRML

► Bentley

Product	Exporter	Native	Properties	Other
AutoPLANT In order to read an AutoPLANT file into JetStream, you either need AutoPLANT installed on the same machine as JetStream, or you will need to install the AutoPLANT Object Enabler. If AutoPLANT is installed, the object enabler is not necessary. With either AutoPLANT or the Object Enabler installed, simply open the AutoPLANT .dwg file in JetStream. You can also export to .nwc or .nwd depending on whether you have a JetStream Roamer or a JetStream Roamer + JetStream Publisher license respectively. JetStream supports AutoPLANT object properties which are read directly from an external database (.mdb). If the .mdb file has the same filename and is located in the same directory as the .dwg, then Equipment, Nozzle and Piping properties will be automatically read (you will need to select these options under Tools > Global Options > DataTools, to enable the automatic reading of these properties). AutoPLANT Central Server databases are also supported. Under Tools > Global Options > DataTools, you can edit the AutoPLANT Central Single-Project (or Multi-Project) link, click on Setup and browse to your AutoPLANT project folder (usually under the C:\Bentley Plant Projects directory). Once this is done, JetStream will automatically read the AutoPLANT properties from the appropriate database. AutoPLANT properties can be viewed in the 'Properties' window. To open the Properties window, go to View > Control Bars > Properties.	✔	✔	✔	DWG DXF 3DS

Continues

Product	Exporter	Native	Properties	Other
MicroStation SE	✔	✔	✔	DWG DXF DGN IGES VRML

Models can be exported from MicroStation SE as either .nwc or .nwd format, depending on whether you have a JetStream Roamer, or JetStream Roamer + Publisher license respectively.

JetStream also supports the native MicroStation .dgn file format, which can be read directly into JetStream.

Object properties, such as colors and ambient, diffuse and shininess properties of materials from .pal and .mat palette and material files are supported. Family and Part information from TriForma and PDS object information can be read from .drv files and DMRS and database linkage and association ID's are also supported. These can be viewed in the Properties window. To open the Properties window, go to View > Control Bars > Properties.

For MicroStation materials and textures to be visible in JetStream, you will require a minimum of a JetStream Roamer + Presenter license.

NOTE: For optimum file fidelity we recommend the NWC (or NWD) Exporter from within MicroStation whenever possible. For exporting numerous files from MicroStation we recommend using the Perl script installed with JetStream (by default in, C:\Program Files\ Navisworks 5\Tools).

Product	Exporter	Native	Properties	Other
MicroStation J	✔	✔	✔	DWG DXF DGN IGES STEP VRML

Models can be exported from MicroStation J as either .nwc or .nwd format, depending on whether you have a JetStream Roamer, or JetStream Roamer + Publisher license respectively.

JetStream also supports the native MicroStation .dgn file format, which can be read directly into JetStream.

Object properties, such as colors and ambient, diffuse and shininess properties of materials from .pal and .mat palette and material files are supported. Family and Part information from TriForma and PDS object information can be read from .drv files and DMRS and database linkage and association ID's are also supported. These can be viewed in the Properties window. To open the Properties window, go to View > Control Bars > Properties.

For MicroStation materials and textures to be visible in JetStream, you will require a minimum of a JetStream Roamer + Presenter license.

NOTE: For optimum file fidelity we recommend the NWC (or NWD) Exporter from within MicroStation whenever possible. For exporting numerous files from MicroStation we recommend using the Perl script installed with JetStream (by default in, C:\Program Files\ Navisworks 5\Tools).

Continues

Product	Exporter	Native	Properties	Other
MicroStation v8 to 8.5	✔	✔	✔	DWG DXF DGN IGES STEP VRML
If you have MicroStation v8 installed on the same machine as JetStream, you can export an .nwc or.nwd, from MicroStation v8, depending on whether you have a JetStream Roamer, or JetStream Roamer + Publisher license respectively.				
JetStream additionally supports v8 DGN files which can therefore be read directly into JetStream.				
Object properties, such as colors and ambient, diffuse and shininess properties of materials from .pal and .mat palette and material files are supported. Family and Part information from TriForma and PDS object information can be read from .drv files and DMRS and database linkage and association ID's are also supported. These can be viewed in the Properties window. To open the Properties window, go to View > Control Bars > Properties.				
For MicroStation materials and textures to be visible in JetStream, you will require a minimum of a JetStream Roamer + Presenter license.				
NOTE: For optimum file fidelity we recommend the NWC (or NWD) Exporter from within MicroStation whenever possible. For exporting numerous files from MicroStation we recommend using the Perl script installed with JetStream (by default in, C:\Program Files\ Navisworks 5\Tools).				
Triforma J	✔	✔	✔	DWG DXF DGN IGES STEP VRML
Models can be exported from MicroStation TriForma J as either .nwc or .nwd format, depending on whether you have a JetStream Roamer, or JetStream Roamer + Publisher license respectively.				
JetStream also supports the native MicroStation TriForma J .dgn file format, which can be read directly into JetStream.				
Object properties, such as colors and ambient, diffuse and shininess properties of materials from .pal and .mat palette and material files are supported. Family and Part information from TriForma and PDS object information can be read from .drv files and DMRS and database linkage and association ID's are also supported. These can be viewed in the Properties window. To open the Properties window, go to View > Control Bars > Properties.				
For MicroStation materials and textures to be visible in JetStream, you will require a minimum of a JetStream Roamer + Presenter license.				
NOTE: For optimum file fidelity we recommend the NWC (or NWD) Exporter from within MicroStation whenever possible. For exporting numerous files from MicroStation we recommend using the Perl script installed with JetStream (by default in, C:\Program Files\ Navisworks 5\Tools).				

Continues

► AVEVA

Product	Exporter	Native	Properties	Other
PDMS		✔		RVM

JetStream can read .rvm files exported from PDMS.

AVEVA Review .rvs files are also supported. These require the same filename as the .rvm file.

PDMS attribute files are supported, output using either the 'Dump Attributes' or 'Datal' formats. These can be viewed in the 'Properties' window. To open the Properties window, go to View > Control Bars > Properties.

NOTE: JetStream Roamer will require an additional JetStream RVM Reader license in order to open .rvm and .rvs file types.

► CADopia

Product	Exporter	Native	Properties	Other
IntelliCAD		✔		DWG DXF

JetStream supports the native IntelliCAD .dwg file format, which can be read directly into JetStream.

► CEA Technology

Product	Exporter	Native	Properties	Other
Plant-4D				DWG DGN

JetStream has file readers for .dwg and .dgn files, both of which are supported by Plant-4D.

► COADE, Inc.

Product	Exporter	Native	Properties	Other
CADWorx Plant	✔	✔	✔	DWG DXF 3DS

Models can be exported from CADWorx as either .nwc or .nwd format, depending on whether you have a JetStream Roamer, or JetStream Roamer + Publisher license respectively.

JetStream also provides support for .dwg files enabling native CADWorx models to be read directly into JetStream.

CADWorx object properties are also supported by Navisworks JetStream and these can be viewed in the 'Properties' window. To open the Properties window, go to View > Control Bars > Properties.

CADWorx piping, steel, and equipment data is fully supported by JetStream.

Continues

Product	Exporter	Native	Properties	Other
CADWorx Pipe	✔	✔	✔	DWG DXF 3DS
Models can be exported from CADWorx as either .nwc or .nwd format, depending on whether you have a JetStream Roamer, or JetStream Roamer + Publisher license respectively.				
JetStream also provides support for .dwg files enabling native CADWorx models to be read directly into JetStream. CADWorx object properties are also supported by Navisworks JetStream and these can be viewed in the 'Properties' window. To open the Properties window, go to View > Control Bars > Properties.				
The legacy product, CADWorx Pipe, is fully supported by JetStream.				
CADWorx Steel	✔	✔	✔	DWG DXF 3DS
Models can be exported from CADWorx as either .nwc or .nwd format, depending on whether you have a JetStream Roamer, or JetStream Roamer + Publisher license respectively.				
JetStream also provides support for .dwg files enabling native CADWorx models to be read directly into JetStream. CADWorx object properties are also supported by Navisworks JetStream and these can be viewed in the 'Properties' window. To open the Properties window, go to View > Control Bars > Properties.				
CADWorx Steel is fully supported by JetStream.				

► COINS

Product	Exporter	Native	Properties	Other
BSLink	✔	✔	✔	DWG DXF 3DS
In order to read a BSLink file into JetStream, you either need BSLink installed on the same machine as JetStream, or AutoCAD installed complete with the BSLink object enabler. If BSLink is installed, the object enabler is not necessary. Simply export to .nwc or .nwd depending on whether you have a JetStream Roamer, or a JetStream Roamer + JetStream Publisher license respectively.				
In JetStream, object properties can be viewed in the 'BSLink' tab in the 'Properties' window. To open the properties window, go to View > Control Bars > Properties.				
Framing	✔	✔	✔	DWG DXF 3DS
In order to read a Framing file into JetStream, you either need COINS Framing installed on the same machine as JetStream, or ADT installed complete with the Framing object enabler				
If COINS Framing is installed, the object enabler is not necessary. Simply export to .nwc or .nwd depending on whether you have a JetStream Roamer, or a JetStream Roamer + Publisher license respectively.				
In JetStream, object properties can be viewed in the Framing tab in the Properties window. To open the properties window, go to View > Control Bars > Properties.				

► CSC

Product	Exporter	Native	Properties	Other
3D+	✔	✔	✔	DWG
3D+ has a Navisworks Exporter, enabling you to export a Navisworks file out of 3D+. Also see the 3D+ website for information on their 3D+ Navisworks Reader. This is a plug-in to JetStream, enabling you to read the native .3nwc file format in JetStream. 3D+ object properties are supported by JetStream via either of the above routes.				

► Dassault Systèmes

Product	Exporter	Native	Properties	Other
CATIA				DXF
JetStream has file readers for .dxf, .igs and .stp files, all of which can be exported from CATIA.				IGES STEP

► Google

Product	Exporter	Native	Properties	Other
SketchUp		✔		SKP
The native SketchUp format, .skp can be read directly into JetStream v5. Texture materials are supported; however, JetStream Roamer will require an additional JetStream Presenter license. Grouping objects in SketchUp will make it easier to use the model in JetStream, for example applying Presenter materials, or attaching objects to TimeLiner tasks.				DWG 3DS

► Hannappel SOFTWARE GmbH

Product	Exporter	Native	Properties	Other
elcoCAD R4	✔	✔	✔	DWG
Models can be exported from elcoCAD as either .nwc or .nwd format, depending on whether you have a JetStream Roamer, or JetStream Roamer + Publisher license respectively. JetStream also supports the native elcoCAD .dwg file format, which can be read directly into JetStream. Object properties are also supported by JetStream and these can be viewed in the 'Properties' window. To open the Properties window, go to View > Control Bars > Properties.				DXF 3DS

► Intergraph

Product	Exporter	Native	Properties	Other
PDS	✔	✔	✔	DWG
				DXF
				DGN
				IGES
				STEP

Intergraph PDS will save the model geometry in a .dgn file and the property information in a .drv file. Both files are required by JetStream to read the model with object properties. In JetStream, go to Tools > Global Options > DGN, scroll down and check the 'Convert PDS Data' option. Ensure the .dgn and .drv files reside in the same directory, then JetStream will import the model, along with its properties.

PDS object properties can be viewed in the 'Properties' window. To open the Properties window, go to View > Control Bars > Properties.

If JetStream has been installed onto the same machine as Intergraph PDS, you will be able to export from PDS as a Navisworks .nwc or .nwd file, depending on whether you have a JetStream Roamer, or a JetStream Roamer + Publisher license respectively.

► Informatix

Product	Exporter	Native	Properties	Other
MicroGDS		✔		MAN

JetStream supports the native MicroGDS .man file format, which can be read directly into JetStream.

Colors and Materials from the file are supported. The .man File Reader options can be found in Tools > Global Options > MAN.

► ITandFactory

Product	Exporter	Native	Properties	Other
TRICAD	✔	✔	✔	DGN
				VRML

JetStream can directly read TRICAD .dgn files, utilizing the .dgn file reader.
If TriCAD is installed on the same machine as JetStream, then you can also export to .nwc or .nwd depending on whether you have a JetStream Roamer, or a JetStream Roamer + Publisher license respectively.

Exporting a Navisworks file from TRICAD also retains the models object properties that can be viewed in the 'Properties' window. To open the Properties window, go to View > Control Bars > Properties.

▶ Kiwi Software GmbH

Product	Exporter	Native	Properties	Other
ProSteel 3D	✔	✔		DWG
This information relates to ProSteel 3D / ProStahl 3D / AutoPLANT Structural.				DXF
In order to read ProSteel files into JetStream, you either need ProSteel 3D installed on the same machine as JetStream, or AutoCAD/MDT/ADT installed complete with the relevant ProSteel 3D object enabler.				3DS
If ProSteel is installed, the object enabler is not necessary. Simply export to .nwd or .nwc depending on whether you have a JetStream Roamer, or JetStream Roamer + Publisher license respectively				
Only basic AutoCAD properties are supported (such as Entity Handle and Material) and these can be viewed in the Properties window. This can be opened by going to View > Control Bars > Properties.				

▶ Kubotek USA

Product	Exporter	Native	Properties	Other
CADKEY				DWG
JetStream has file readers for .dwg files and also .dxf, .igs and .stp files, all of which are supported by CADKEY.				DXF
				IGES
				STEP

▶ M.A.P.

Product	Exporter	Native	Properties	Other
CAD-Duct	✔	✔	✔	DWG
In order to read a CAD-Duct file into JetStream, you either need CAD-Duct installed on the same machine as JetStream, or have the CAD-Duct Object Enabler installed. The CAD-Duct Object Enabler can be downloaded from the CAD-Duct website, found under Support > Additional Files.				DXF
				3DS
If CAD-Duct is installed, the Object Enabler is not necessary. Simply export to .nwd or .nwc depending on whether you have a JetStream Roamer, or JetStream Roamer + JetStream Publisher license respectively.				
In JetStream, object properties can be viewed in the 'CAD-Duct' tab in the 'Properties' window. To open the Properties window, go to View > Control Bars > Properties.				

▶ McNeel North America

Product	Exporter	Native	Properties	Other
Rhino				DWG
To view your Rhino model in JetStream, you will have to export to one of the supported file formats. This can then be read directly into JetStream.				DXF
				3DS
				IGES
				STEP

► **Mensch und Maschine**

Product	Exporter	Native	Properties	Other
RoCAD	✔	✔	✔	DWG DXF 3DS
Models can be exported from RoCAD as either .nwc or .nwd format, depending on whether you have a JetStream Roamer, or JetStream Roamer + Publisher license respectively.				
JetStream also supports the .dwg file format, which can be read directly into JetStream.				
If you do not have RoCAD installed on the same machine as JetStream, you will need to have AutoCAD installed, along with the RoCAD Object Enabler.				
Object properties, (such as Material Colors and Entity Handles) are also supported by JetStream and these can be viewed in the 'Properties' window. To open the Properties window, go to View > Control Bars > Properties.				

► **MultiSUITE**

Product	Exporter	Native	Properties	Other
MultiSTEEL	✔	✔	✔	DWG DXF 3DS
.dwg files from MultiSteel can be read directly into JetStream utilizing the .dwg file reader.				
Models can also be exported from MultiSteel as either .nwc or .nwd format, depending on whether you have a JetStream Roamer, or JetStream Roamer + Publisher license respectively.				
MultiSteel object properties can be viewed in the 'Properties' window. To open the Properties window, go to View > Control Bars > Properties.				

► **Nemetschek**

Product	Exporter	Native	Properties	Other
Allplan				DWG DXF DGN IFC
All Plan files can be viewed in JetStream if they have been saved in one of the above formats.				
Object properties are not supported.				

► **PROCAD**

Product	Exporter	Native	Properties	Other
3DSMART	✔	✔		DWG DXF 3DS
With 3DSMART installed on the same machine as JetStream, you can read your DWG file directly into JetStream, utilizing the DWG file reader.				
You can also export an .nwc or .nwd file directly from 3DSMART, depending on whether you have a JetStream Roamer, or JetStream Roamer + JetStream Publisher license respectively.				

▶ PTC

Product	Exporter	Native	Properties	Other
Pro/ENGINEER You can read your Pro/ENGINEER models into JetStream via any of the file formats listed.				IGES STEP VRML
CADDS 5 You can read CADDS 5 3D models into JetStream, by exporting either an IGES or STEP file from CADDS 5 and then utilize either the JetStream .igs, or .stp file readers.				IGES STEP VRML

▶ QuickPen

Product	Exporter	Native	Properties	Other
PipeDesigner 3D Models can be exported from PipeDesigner 3D as either .nwc or .nwd format, depending on whether you have a JetStream Roamer, or JetStream Roamer + Publisher license respectively. JetStream also supports the native .dwg file format, which can be read directly into JetStream. Object properties, (such as Material Colors and Entity Handles) are also supported by JetStream and these can be viewed in the 'Properties' window. To open the Properties window, go to View > Control Bars > Properties.	✔	✔	✔	DWG DXF 3DS
DuctDesigner 3D Models can be exported from DuctDesigner 3D as either .nwc or .nwd format, depending on whether you have a JetStream Roamer, or JetStream Roamer + Publisher license respectively. JetStream also supports the native .dwg file format, which can be read directly into JetStream. Object properties, (such as Material Colors and Entity Handles) are also supported by JetStream and these can be viewed in the 'Properties' window. To open the Properties window, go to View > Control Bars > Properties.	✔	✔	✔	DWG DXF 3DS

▶ RAM International

Product	Exporter	Native	Properties	Other
CADstudio Models can be exported from CADstudio as either .nwc or .nwd format, depending on whether you have a JetStream Roamer, or JetStream Roamer + JetStream Publisher license respectively. JetStream also supports the native CADstudio .dwg file format, which can be read directly into JetStream. Object properties are also supported by JetStream and these can be viewed in the 'Properties' window. To open the Properties window, go to View > Control Bars > Properties.	✔	✔	✔	DWG DXF 3DS

► **SolidWorks**

Product	Exporter	Native	Properties	Other
SolidWorks 97+ to 2007 SolidWorks part and assembly files can be read directly into JetStream, utilizing the SolidWorks file reader.		✔		SLDASM SLDPRT IGES STEP VRML

► **Tekla**

Product	Exporter	Native	Properties	Other
Tekla Structures Tekla Structures can export to the .dgn file format, which can be read directly into JetStream. Tekla Structures can additionally output to CIS/2. This may then be converted to VRML using the translator from the National Institute of Standards and Technology (see the NIST website for more information). The VRML file can then be read directly into JetStream. Limited object properties are supported via the .dgn route.			✔	DGN VRML
Xsteel Xsteel can export to the .dgn file format, which can be read directly into JetStream. Limited object properties are supported via the .dgn route.			✔	DGN

► **think3**

Product	Exporter	Native	Properties	Other
Thinkdesign JetStream has file readers for .dwg/.dxf files and also .igs and .stp files.				DWG DXF IGES STEP

► **UGS**

Product	Exporter	Native	Properties	Other
I-deas JetStream has file readers for .dxf, .igs and .stp files.				DXF IGES STEP
Solid Edge 19 (and earlier) JetStream supports native Solid Edge files, allowing them to be read directly into JetStream. File readers for .igs and .stp files are also available.		✔		ASM PAR PSM IGES STEP

Continues

Product	Exporter	Native	Properties	Other
NX (Unigraphics) JetStream has file readers for .dxf, .igs and .stp files.		✔		DXF IGES STEP
FactoryCAD In order to read a FactoryCAD file into JetStream, you either need FactoryCAD or the FactoryCAD object enabler installed on the same machine. If FactoryCAD is installed, the object enabler is not necessary. Simply export to .nwc or .nwd depending on whether you have a JetStream Roamer, or a JetStream Roamer + Publisher license respectively.	✔	✔		DWG DXF 3DS

► UHP Process Piping

Product	Exporter	Native	Properties	Other
FabPro Pipe Models can be exported from FabPro Pipe as either .nwc or .nwd format, depending on whether you have a JetStream Roamer, or JetStream Roamer + Publisher license respectively. JetStream also supports the native FabPro Pipe .dwg file format, which can be read directly into JetStream.	✔	✔		DWG DXF 3DS

► x-plant

Product	Exporter	Native	Properties	Other
x-plant Models can be exported from x-plant as either .nwc or .nwd format, depending on whether you have a JetStream Roamer, or JetStream Roamer + Publisher license respectively. JetStream also supports the native x-plant .dwg file format, which can be read directly into JetStream.	✔	✔		DWG

Index

B